Wernfried Heilen

Sascha Herrwerth

Silicone Resins and their Combinations

Translated and edited by John Haim and David Hyatt

W. Heilen, S. Herrwerth: Silicone Resins and their Combinations

ISBN: 978-3-86630-828-2

Cover: lunamaria/Fotolia.com

Bibliographische Information der Deutschen Bibliothek
Die Deutsche Bibliothek verzeichnet diese Publikation in der Deutschen Nationalbibliographie; detaillierte bibliographische Daten sind im Internet über http://dnb.ddb.de abrufbar.

Wernfried Heilen and Sascha Herrwerth
Silicone Resins and their Combinations
Hanover: Vincentz Network, 2015
European Coatings Library
ISBN 3-86630-828-0
ISBN 978-3-86630-828-2

Please ask for our book catalogue
Vincentz Network, Plathnerstr. 4c, 30175 Hanover, Germany
T +49 511 9910-033, F +49 511 9910-029
books@european-coatings.com, www.european-coatings.com

Layout: Vincentz Network, Hanover, Germany
Printed by: Quensen Druck + Verlag GmbH & Co. KG, Hildesheim, Germany

ISBN 3-86630-828-0
ISBN 978-3-86630-828-2

European Coatings // LIBRARY

Wernfried Heilen

Sascha Herrwerth

Silicone Resins and their Combinations

Translated and edited by John Haim and David Hyatt

W. Heilen, S. Herrwerth: Silicone Resins and their Combinations

ISBN: 978-3-86630-828-2

Foreword

Because of their molecular structure, silicone resins and silicone combination resins are used in numerous industrial applications, particularly as binders, for formulating coatings. The linking of silicon and oxygen atoms to form a stable basic framework, in which the free valencies of the silicon are saturated by hydrocarbon groups, results in outstanding properties which cannot be achieved with other products. The many possible combinations of the silicone building blocks are reflected in the impressive diversity of silicone chemistry and the resultant products.

The present book (2nd edition) is intended to provide a concentrated overview of the chemistry and technology of silicone resins and their use from an industrial viewpoint.

It aims to report on current developments in the field of silicones for coatings and gives those approaching the subject an overview of the most important areas of application.

Since publication of the 1st edition of this book almost ten years ago, some areas of application have seen further technological developments and these have driven significant advances.

For example, silicone combination resins are being increasingly used in anti-corrosion coatings to cope with extreme weather conditions and temperatures.

Alkoxy-silyl functional urethane resins are used in high-tech coatings to increase scratch resistance.

New types of acrylic dispersions in combination with functional polysiloxanes enable the formulation of "below-critical" emulsion coatings with similar properties to "above-critical" silicone resin coatings.

These and further innovations are described in the 2nd edition of this book and reflect the current state of the art in this field.

Since this field will continue to develop, I am pleased that I have been able to win my colleague Dr. Sascha Herrwerth to join me as co-author for this new edition.

I would like to thank our colleagues Dr. Michael Ferenz, Dirk Hinzmann and Dr. Berendjan de Gans for numerous technical discussions.

Thanks are also due to Evonik for making available company literature and permission to reprint extracts as well as for technical and material resources.

Wernfried Heilen

Essen, July 2014

Contents

W. Heilen, S. Herrwerth: Silicone Resins and their Combinations

ISBN: 978-3-86630-828-2

1 Introduction

The outstanding properties of silicone resins and silicone-containing combination resins in the form of solutions, liquid resins and emulsions make them extremely versatile. The versatility of silicone resins stems largely from the fact that the organo-siloxanes, on which they are based, can be modified in many ways and combined with numerous organic polymers and different functionalities. Thus the unique properties of the organo-siloxanes, such as high surface activity, chemical inertness and thermal stability can be combined with the properties of the chosen organic functionality or polymer. Applications range from impregnations and weather-resistant exterior coatings for building conservation, through heavy duty anti-corrosion protection to high temperature resistant coatings. Acrylic-functional silicone resins and epoxy-functional silicone resins are reactive resins, which are used as UV-curable coatings for paper and as additives for radiation curing printing inks and wood lacquers.

1.1 Organo-siloxanes and organo-polysiloxanes

The starting products required to manufacture silicones, R_3SiOH (silanols), $R_2Si(OH)_2$ (silane diols), $RSi(OH)_3$ (silane triols) and $Si(OH)_4$, are obtained by hydrolysis of the corresponding halogen compounds R_3SiCl, R_2SiCl_2, $RSiCl_3$ and $SiCl_4$[1]. The latter can be prepared by addition of alkyl halides RX to very pure silicon (> 98 %) in the presence of a copper catalyst[2].

This industrial process today generally known as Müller-Rochow Synthesis is also termed "direct synthesis" since organo-chlorosilanes are manufactured directly from elemental silicon (Figure 1.1). This process was discovered by

$$Si + RCl \xrightarrow[\text{Cat.}]{300 - 400\ °C} R_xSiCl_{4-x}$$

Cat. = Cu

R = Alkyl, Aryl

Figure 1.1: Organo-chlorosilanes manufactured by Müller-Rochow-Synthesis

W. Heilen, S. Herrwerth: Silicone Resins and their Combinations

ISBN: 978-3-86630-828-2

Müller and Rochow in the early 1940's and was the starting point for silane and silicone chemistry to grow and become economically viable on a large scale.

Table 1.1: Typical product distribution for the reaction of silicon with methyl chloride (Müller-Rochow-Synthesis)

Product	Parts by weight [%]
R_2SiCl_2	70 to 90
$RSiCl_3$	5 to 15
R_3SiCl	2 to 4
RSiHCl2	1 to 4
$R_nSi_2Cl_{6-n}$	3 to 8

The various by-products formed in this reaction can be separated by distillation. The most important products by volume are methy-chlorosilanes produced by "Direct Synthesis" with methyl chloride and elemental silicon[3]. Longer chain alkyl halides result in low yields of product. Table 1.1 shows a typical product distribution for the reaction of silicon with methyl chloride.

Alternatively the halogen compounds can be prepared by alkylating silicon tetrachloride with alkylating agents such as ZnR_2, AlR_3, MgRCl (R = alkyl or aryl group).

Although the Grignard process is more complex than direct synthesis, it permits, for example, manufacture of mixed aryl/alkylchlorosilanes, i.e. silanes with different organic groups at the same silicon atom[4]. This can be of interest where the polysiloxane end product is required to have specific properties. Besides direct synthesis, the Grignard process investigated by Kipping[5] is thus still of considerable industrial interest (Figure 1.2).

Hydrolysis of dichlorodialkyl silanes R_2SiCl_2 leads to dihydroxydialkyl silanes $R_2Si(OH)_2$. However these are not stable and condense immediately to polymeric silicones $(R_2SiO)_n$ with the elimination of water.

The ratio of cyclic to linear compounds and the chain length of the linear siloxanes as well as the properties of the resultant polymer can be controlled by the hydrolysis conditions: basic catalysts and high temperatures favour the formation of high molecular weight linear polymers. Acidic catalysts result in the formation of low molecular weight polymers and cyclic oligomers.

Figure 1.2: Manufacture of mixed aryl/alkylchlorosilanes by the Grignard synthesis

$$2\,RMgCl + SiCl_4 \rightarrow R_2SiCl_2 + 2\,MgCl_2$$

and

$$R'MgCl + R_2SiCl_2 \rightarrow R_2R'SiCl + MgCl_2 \text{ etc.}$$

R = Alkyl
R' = Aryl

A large proportion of polysiloxanes is currently manufactured by ring-opening polymerization (equilibration). For this, cyclic tetramers and pentamers are manufactured from dichlorodimethyl silane with the help of ionic initiators. The cyclic oligomers are separated from the reaction mixture and used for the ring-opening polymerization.

The equilibration can be acid or alkaline catalysed. After deactivation of the catalyst, 10 to 20 % of cyclic siloxanes, mostly tetramers and pentamers, can be removed by vacuum distillation. If the catalyst is not deactivated, the removed cyclic siloxanes are permanently regenerated until the linear polymers are completely used up.[6]

Since the basic composition R_2SiO corresponds with the formula of organic ketones R_2CO, they were called silico-ketones (silicones) by their discoverer (F. S. Kipping), a name which was then extended to the entire class of organo-silicone oxygen compounds.

The term siloxane, derived from the expression Sil-Oxan for the Si-O-Si-bond, is however more accurate. If one or more organic groups, such as methyl, phenyl, octyl or aminoalkyl, are attached to each silicon atom, the compounds are called organo-siloxanes. Such monomers are used as building blocks to manufacture silicone polymers, i.e. organo-polysiloxanes (Table 1.2).

Table 1.2: Building blocks for polysiloxanes

Monomere (silanes)	Organo-siloxane	Symbol
$(R)_3SiCl$	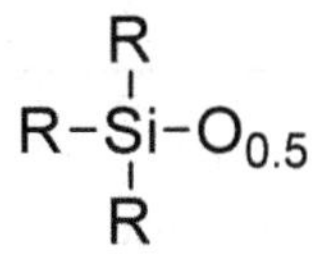	M
$(R)_2SiCl_2$	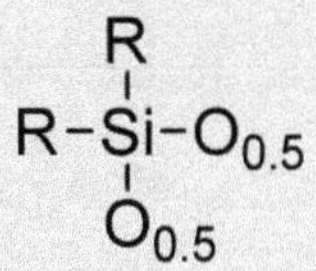	D
$RSiCl_3$	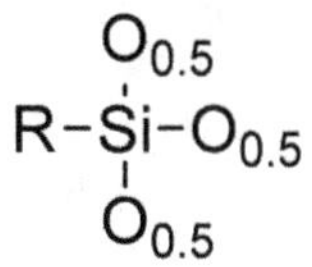	T
$SiCl_4$	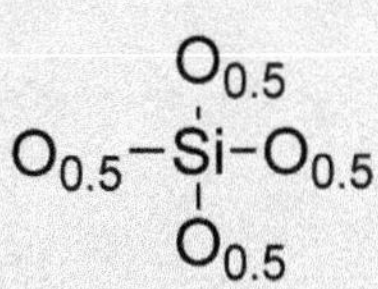	Q

The fact that various siloxane units in the molecule can be combined with each other leads to a huge diversity of compound types[1]. The following are examples of the compounds which have been synthesized:

1. A monofunctional siloxane unit reacts once with a similar unit resulting in a hexaorgano-disiloxane:

$$R_3Si{-}O{-}SiR_3 \qquad [M_2]$$

Figure 1.3

2. Closed rings result when difunctional units combine. The smallest known ring contains three siloxane units; rings with four or five siloxane units are easiest to create (Figure 1.4).

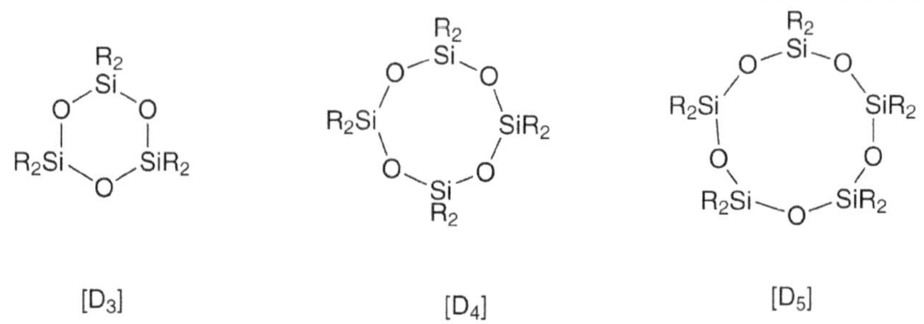

Figure 1.4

3. Trifunctional siloxane units linked together generally yield randomly, three-dimensional crosslinked molecules (Figure 1.5).

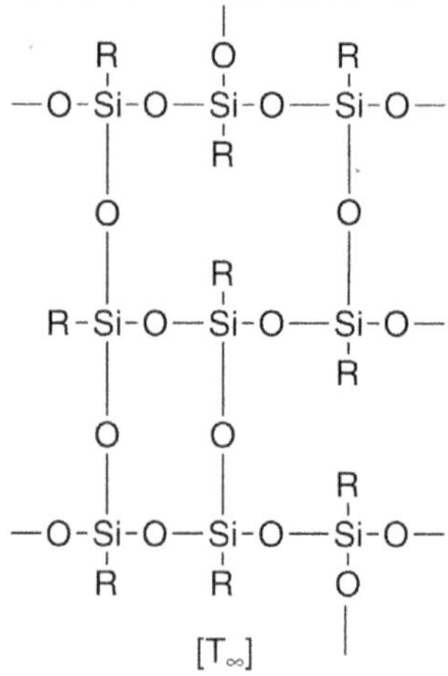

Figure 1.5

Under certain conditions, however, small cage-like structures with four, six and twelve siloxane units, which can be considered polycyclic, have been found.

4. A combination of mono and difunctional siloxane units leads to linear polymeric siloxanes with a huge variety of chain lengths determined by the ratio of di to monofunctional units (Figure 1.6).

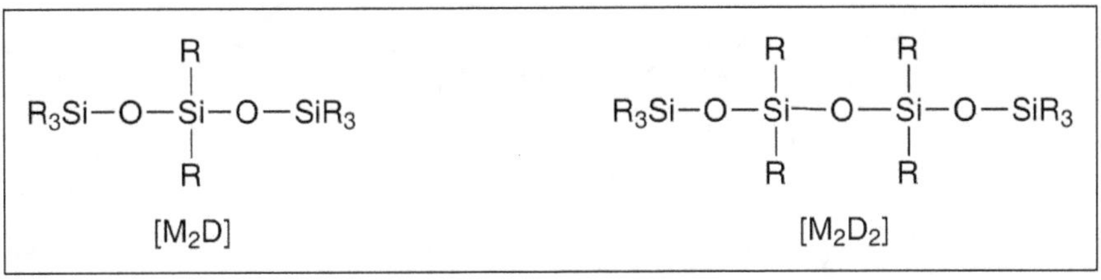

Figure 1.6

Similar processes can lead to linear high polymers (Figure 1.7).

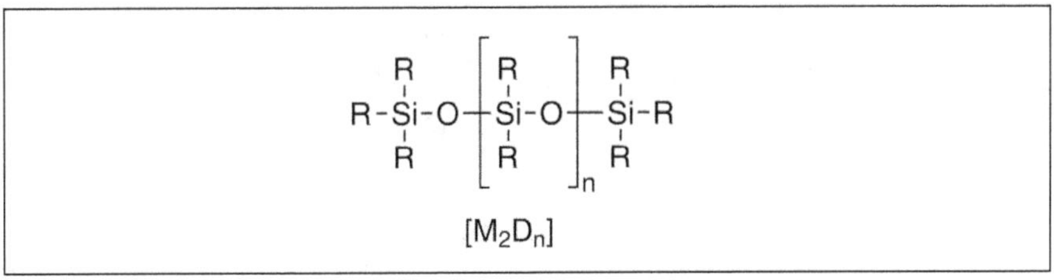

Figure 1.7

5. The combination of mono- and tri- or tetrafunctional siloxane units results in low molecular weight structures such as shown in Figure 1.8.

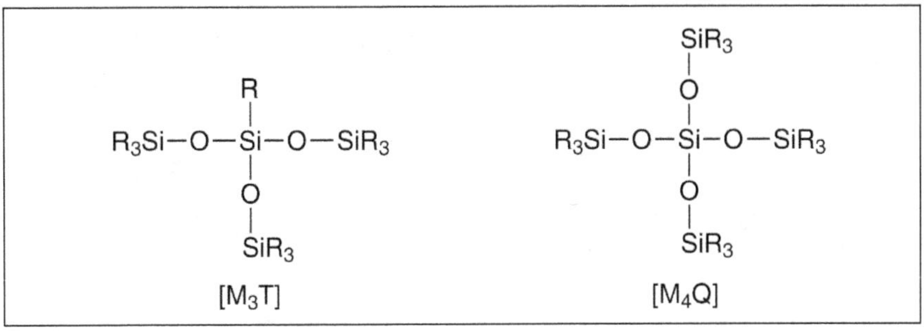

Figure 1.8

6. Linking di and trifunctional siloxane units generally leads to macromolecules, which are usually networks, if they have a high T-unit content (Figure 1.9) while an excess of D-units results in chains with a low degree of crosslinkage (Figure 1.10).

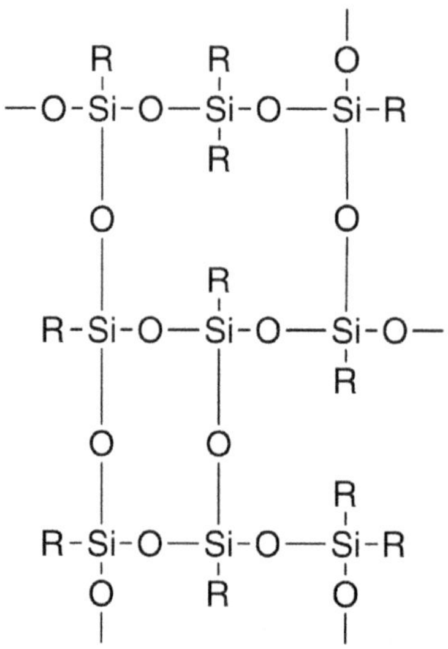

$[D_nT_m(m > n)]$

Figure 1.9

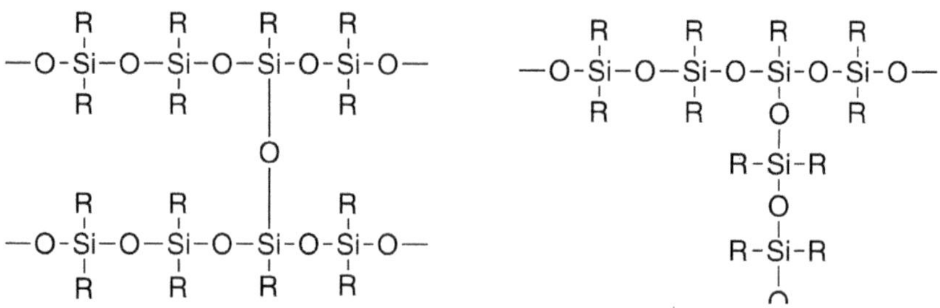

$[D_nT_m(n >> m)]$ $[D_nT_m(n >> m)]$

Figure 1.10

On the other hand, D- and T-units can congregate in limited numbers to form low molecular weight structures (Figures 1.11):

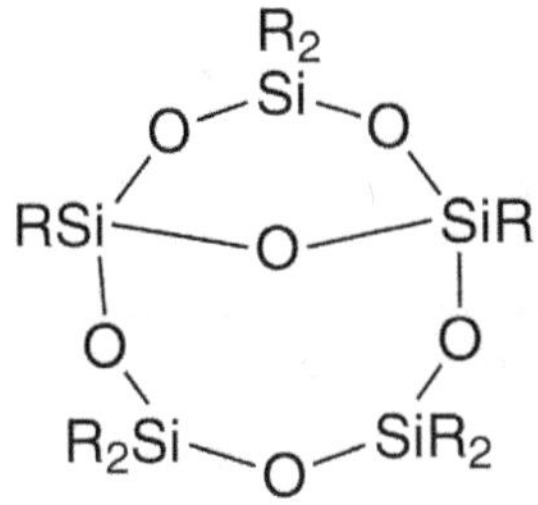

$[D_3T_2]$

Figure 1.11

7. Finally, the reaction of di- and tetrafunctional units yields not only strongly, crosslinked groups of molecules (as with the combination of di and trifunctional components) but, under specific conditions, spiro compounds (Figure 1.12):

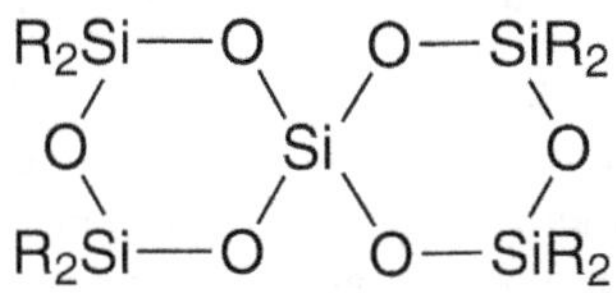

$[D_4Q]$

Figure 1.12

In line with their functionality, the D-unit is used as a chain or ring former, the M-unit as a stopper and the T- and Q-units as crosslinkers.

1.2 Chlorosilanes – the building blocks for silicone resins

Silicone resins are highly crosslinked siloxane systems. The crosslinking components are introduced with tri- or tetra-functional silanes. Only a few silanes have attained practical importance as resin building blocks and can be easily manufactured on a large scale. These are: methyl-trichlorosilane, phenyl-trichlorosilane, dimethyl-dichlorosilane, phenylmethyl-dichlorosilane, diphenyl-dichlorosilane, trimethyl-chlorosilane and tetrachlorosilane[7]. Methyl-trichlorosilane is the most important monomer by volume in silicone resin technology. Normally combinations of various silanes are used in the synthesis of silicone resins[8].

The choice of chlorosilanes for a particular silicone resin determines its characteristics. Prediction of specific resin properties as a function of composition frequently fails since processing and curing conditions influence the final molecular configuration and related characteristics.

However, some generalizations can be made:

Methyl-trichlorosilane leads to high hardness and rapid curing but also brittleness, poor pigmentability and poor compatibility with organic resins. This is due to the fact that the methyl-trichlorosilane contains the smallest amount of carbon of all chlorosilanes, except for the less used tetrachlorosilane, and is much closer to inorganic silicates than, for example, dimethyl-dichlorosilane or phenyl-trichlorosilane.

The use of phenyl-trichlorosilane results in resins with high thermal stability, good pigmentability, increased compatibility with organic resins but in a high degree of crosslinking and comparatively brittle products of low thermoplasticity as the methyl-trichlorosilane. In practice phenyl-trichlorosilane is not used alone.

Flexibility is achieved primarily by the concomitant use of dimethyl-dichlorosilane but at the cost of sacrificing hardness at elevated temperatures. This applies particularly to dimethyl-dichlorosilane, but also, to a lesser extent, to phenylmethyl-dichlorosilane and diphenyl-dichlorosilane. In the synthesis of silicone resins, cyclic siloxanes, such as D4 and D5, can also be used instead of dichlorosilane derivatives to achieve flexibility.

Table 1.3: Effect of silanes on the properties of silicone resin films

Property	CH_3SiCl_3	$C_6H_5SiCl_3$	$(CH_3)_2SiCl_2$	$CH_3(C_6H_5)SiCl_2$	$(C_6H_5)_2SiCl_2$
Hardness	↗	↗	↘	↘	↘
Tack	↘	↘	↗	↗	↗
Curing speed	↑	↗	↘	↘	↓
Flexibility	↓	↘	↗	↗	↗
Compatibility*	↓	↑	↘	↑	↑
Pigmentability	↓	↑	↘	↑	↑

* with organic resins

Trimethyl-chlorosilane is often used as a monofunctional silane in combination with trichloro- or tetrachlorosilane derivatives to obtain more stable resin end products and a lower degree of crosslinking, which leads to less brittle coatings with lower susceptibility to cracking under thermal stress. Because of its monofunctionality, trimethyl-chlorosilane is used as an end blocking agent in silicone resin synthesis. Reproducible incorporation of trimethylsilyl groups in the polymer frequently poses considerable difficulties because trimethyl-chlorosilane exhibits comparatively low reactivity during hydrolysis/condensation. Hexamethyl-disiloxane can be used instead of trimethyl-chlorosilane, in silicone resin synthesis.

Higher aliphatic groups (usually butyl or octyl) increase compatibility with organic polymers but significantly impair heat resistance.

Aminoaliphatic groups (aminopropyl- or aminoethyl aminopropyl groups) lead to specific properties in water repellency applications for the construction business.

Methylvinyl dichlorosilanes can be used to incorporate reactive vinyl-groups in the silicone resin structure. The vinyl functionality can be used for further crosslinking of the silicone resins by addition curing.

All organo-oligosiloxanes used on a large scale have the general formula (Figure 1.13) and are obtained primarily by hydrolysis of organo-chloro- or organo-alkoxy-siloxanes.

Hydrolysis of chlorosilanes with alcohols under partial condensation leads to oligosiloxanes or low molecular weight alkoxy-functional siloxanes. The degree of condensation n is of major importance.

$$\left[R^1{}_a - \underset{\displaystyle O_{\frac{4-(a+b)}{2}}}{\underset{|}{Si}} - (OR^2)_b \right]_n$$

R^1 = Org. group

R^2 = Hydrogen or alkyl group

$a \leq 1 - 1.5^*$, $b \leq 1$

* $a \geq 1.5$ in case of silicone combination resins

Figure 1.13: General formula for organo-oligosiloxanes

The functional groups OR^2 (where R^2 stands for hydrogen or an alkyl group with one to four C-atoms), obtained by hydrolysis or alcoholysis, contribute decisively to the properties of the silicone resin pre-products. These functional groups are responsible for the reactivity and the stability of the intermediate and silicone resin end products. They also govern the functional properties of the cured coating film.

1.3 Manufacture of the resin intermediates

The first step in preparing silicone resin intermediates consists of formulating an appropriate mixture of organo-chlorosilanes in an aqueous medium to obtain organo-oligosiloxane intermediates by hydrolysis and partial condensation[9].

Almost all industrial silicone resins contain methyl- or phenyl groups as Si-C bonded organic groups R^1 (see Figure 1.13). Other organic groups such as higher alkyl, substituted alkyl- or aryl- but also vinyl-groups are quantitatively unimportant but are used in some resins as modified structural components.

The chlorosilanes used are immiscible with water and the various chlorosilanes hydrolyse differently, leading often to undesired reactions and even to gelation of the products. The blend may therefore be mixed with inert solvents, which serve to modify the rate of hydrolysis and provide a diluent for the hydrolysed resin intermediate. The most frequently used solvents are hexane, heptane, toluene, and xylene and esters such as butyl acetate and chlorinated hydrocarbons.

The manufacture of the resin intermediates is complicated by a number of factors. The hydrolysate tends to gel and become insoluble if the content of tri-functional or tetra-functional chlorosilanes is too high, the solvent concentration is too low, or conditions are not carefully controlled to prevent excessive silanol condensation. Gel formation can be minimized by the addition of modifiers, usually low molecular weight alcohols such as methanol, ethanol or butanol. Besides the dilution effect, the alcohol reacts in the alcoholysis reaction with the Si-Cl groups to yield alkoxysilanes rather than silanols. The alkoxysilyl groups thus formed are considerably less reactive during hydrolysis/condensation than the Si-Cl groups and facilitate control of the reaction. The low molecular alcohols contribute additionally to compatibility between the resin phase and the hydrochloric acid phase (2-phase solvent). Silicone resin intermediates containing alkoxysilyl groups are more stable than silanol functional intermediates and can be used, for example, to advantage in further reactions with OH-functional organic polymers or organic polyols in the conversion to silicone resins.

The difference in the rates of hydrolysis of various chlorosilanes is the second complicating factor. As a rule, under certain conditions, the hydrolysis rate increases with higher chloro-functionality and decreases with increasing molecular weight of the organic substituent groups. In order to obtain the desired final properties, conditions must be balanced to promote incorporation of all the hydrolysate products in the average resin molecule. This can be realized through appropriate selection of the solvents, proper agitation, and, if necessary, sequential addition of the chlorosilane monomers to be hydrolysed.

The silicone resin intermediate synthesis may be carried out batchwise or continuously.

The continuous process has, in theory at least, the advantage of providing better uniformity since each increment of hydrolysate is processed in the same way; the molecular weight distribution should therefore be narrower[10].

There are two different processes for batchwise production, which differ in the order in which the reactants, i.e., the silanes and the hydrolysis medium, are added:

1. Direct hydrolysis; the silane mixture is added to the hydrolysis medium.
2. Reverse hydrolysis; the hydrolysis medium is added to the silane mixture.

The reverse hydrolysis is, however, exceedingly difficult to control, particularly in the case of hydrolysis/condensation of mixtures of trichlorosilanes with dichlorosilanes or monochlorosilanes. With the usual concomitant use of low-boiling methylchlorosilanes, irreproducible silane losses, which disrupt stoichiometry, can occur. Consequently, silicone resins are manufactured nowadays mainly by direct hydrolysis.

In practice, a considerable excess of water is used for hydrolysis, and the chlorosilane solvent mixture is fed-in at a controlled rate. The initial phase of the hydrolysis/condensation reaction can be conducted at room temperature. Following completion of the silane addition, it is desirable to heat the reaction mixture to above room temperature to bring the reaction to completion.

It has already been mentioned that the primary products of silane hydrolysis/alcoholysis taking place under partial condensation are silicone resin intermediates and not silicone resins. They bear many OH and/or OR groups as end groups, but do not contain chlorine. The silicone resin intermediates, which are stable under certain conditions, must be converted into a stable final state before use as resin end products. This additional process step can be carried out immediately after the hydrolysis/condensation step.

The molecular weight of the resin intermediates is usually between 600 and 5000, but oligomer mixtures with larger or smaller molecular weights are known. The molecular weight distribution and the structure of the individual molecules determine subsequent binder properties.

The complex preparation of the silicone resin intermediates and the resulting structure of the silicone resins make it impossible to give a precise chemical structural formula. This is why silicone resin intermediates and silicone resins are conventionally only characterized by their reactive groups (e.g. silanol, methoxy), by their substituents (e.g. methyl, phenyl) and by their molecular weight.

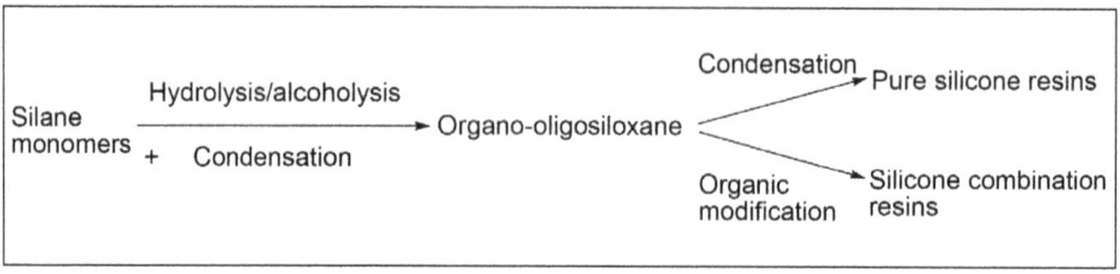

Figure 1.14: Manufacture of silicone- and silicone combination resins

The organo-oligosiloxane intermediates obtained by hydrolysis/alcoholysis and partial condensation of organo-chloro- or organo-alkoxy-siloxanes can be converted to film-forming, curable silicone resins by partial condensation (frequently called "bodying") of silanol groups or to silicone combination resins. In the process, condensation of the intermediates results in pure silicone resins. In contrast, modification with organic resins or organic resin intermediates results in silicone combination resins (Figure 1.14).

During the "bodying" of silicone resin intermediates to silicone resins, the residual silanol groups partially condense and, consequently the molecular weight of the silicone resins increases. This reaction can be promoted by metal catalysts or acid-treated clays. Polyols can also be added as coupling agents during the "bodying" process.

The silicone resin intermediates are classified according to the particular combination of silicone units that make up the intermediate. The most important class of silicone resin intermediates by production volume is the DT silicone resin intermediates composed of diorgano-dichloro and monoorgano-trichlorosilanes. The most important silanes by volume for DT silicone resins are methyl-trichlorosilane, phenyl-trichlorosilane and dimethyl-dichlorosilane.

The R^1/Si ratio (see Figure 1.13) of most pure DT silicone resins is between 1.0 and 1.5. However, this ratio exceeds 1.5 in certain siloxane components in silicone combination resins. The manufacture of DT silicone resin intermediates has frequently been described in the patent literature[11–13].

MQ or MQD resin intermediate processing is similar to that outlined above for DT resins. However, for the MQ silicone resin intermediates, trimethyl-chlorosilane and tetrachlorosilane are the most important silanes and for MQD silicone resin intermediates, trimethyl-chlorosilane, dimethyl-dichlorosilane, methylphenyl-dichlorosilane and tetrachlorosilane are the most important. Production volumes of MQ and MQD resin intermediates and the correlating resins are smaller than that of DT resin intermediates[14]. An alternative to silicon tetrachloride as a starting tetrafunctional material involves processes based on the conversion of sodium silicate to a silicic acid solution, which is then treated with chlorosilanes[15].

2 Silicone resins

2.1 Pure silicone resins

The conversion of silicone resin intermediates (see Chapter 1.3) into the final resin end state can be achieved by condensation of the silanol groups with an increase in molecular weight according to the equation shown in Figure 2.1.

$$\equiv Si-OH + HO-Si\equiv \longrightarrow \equiv Si-O-Si\equiv + H_2O$$

Figure 2.1: Conversion of organo-oligosiloxanols into resin end state through condensation of silanol groups

This relatively slow reaction, which can generally be promoted by metal catalysts, for example organo-tin compounds, or acid-treated clays, is frequently called "bodying" and leads to a relatively broad molecular weight distribution[9]. To achieve resin structures with as narrow and reproducible a molecular weight distribution as possible, the products obtained in this manner can be subjected to a further step of catalytic equilibration, similar to the manufacture of linear silicone fluids[16]. In this process the siloxane bonds are broken and re-established at new positions. The silanol group content of the silicone resin does not change during this equilibration.

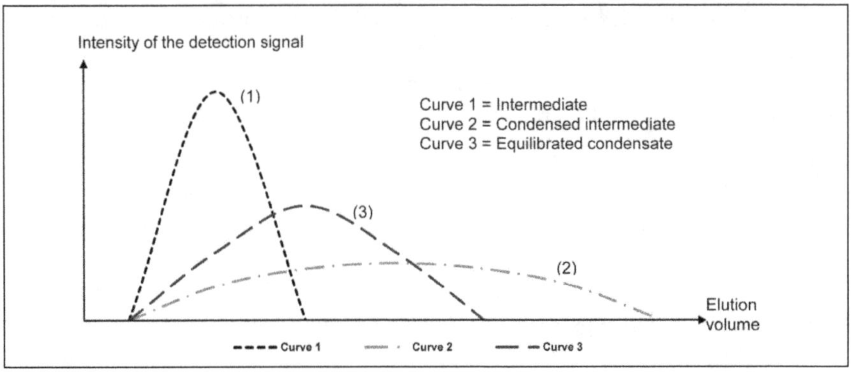

Figure 2.2: Schematic GPC chromatograms for (1) silicone resin intermediate, (2) condensed/bodied silicone resin and (3) equilibrated condensed silicone resins

W. Heilen, S. Herrwerth: Silicone Resins and their Combinations

ISBN: 978-3-86630-828-2

Figure 2.2 shows three schematic GPC-chromatograms[17]. Curve 1 characterizes the low molecular weight resin intermediate, curve 2 the resin after condensation and curve 3 the resin after equilibration. The catalytic equilibration is a standard procedure in the production process of MDQ silicone resins from MQ resins intermediates[9].

An alternative technology for the preparation of silicone resins from the hydroxyl-functional silicone resin intermediates is the condensation of silanol-free alkoxy-functional silicone resin intermediates (see Chapter 1.3). These can be crosslinked with low molecular weight organic polyols, such as glycerol, trimethylol propane, trimethylol ethane or pentaerythritol[18]. This reaction is an ester interchange shown in Figure 2.3 and can be controlled more easily than the condensation of the hydroxyl-functional silicone resin intermediates described in Figure 2.1.

$$\equiv Si-OR + HO-C\equiv \longrightarrow \equiv Si-O-C\equiv + ROH$$

Figure 2.3: Condensation of silanol-free alkoxy-functional silicone resin intermediates with organic polyols

The polyol content in these resins should generally be far below 10 %, so that the high thermal stability application properties of the pure silicone resins are not impaired.

The wide range of monomers and reaction conditions which can be used in the manufacture of silicone resin intermediates and in their subsequent conversion into silicone resins permits a multitude of silicone resin structures to be obtained. The chemical properties and application-related characteristics of these differ and can be optimized to suit requirements. In principle, the various silicone resins can be divided into three classes but the distinction between them is blurred. They are shown schematically in Figure 2.4.

In general the conversion of the silicone resin intermediates into a processable silicone resin is interrupted at a certain degree of the condensation. It is later allowed to run to completion during the film curing of the silicone resin coating. The curing process of the silicone resin coating can be accelerated with suitable catalysts or by forced drying with high temperatures[19]. Complete condensation of all available silanol groups generally does not occur during film drying for steric reasons. The pure silicone resins are often used in heat resistant coatings and, in particular, in high temperature resistant anti-corrosion systems, or as co-binders in physical mixtures (cold blends) with other organic binders to improve, for example, heat or weathering resistance of the organic binders. In cold blends the compatibility of the silicone resins with the organic binders has to be carefully controlled in the formulation and during the curing of the coating.

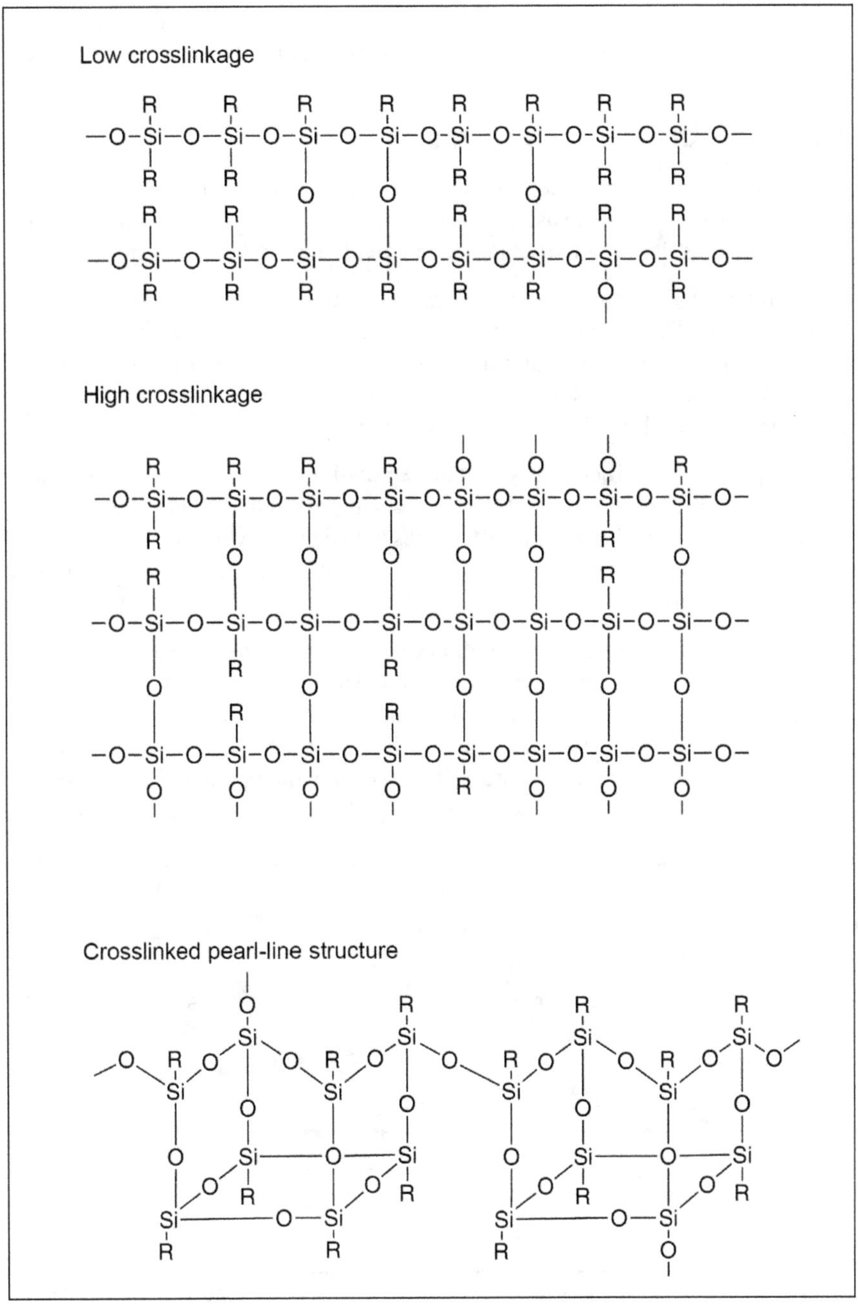

Figure 2.4: Schematic diagram of different silicone resin structures

A broad range of different silicone resins is commercially available from, for example, Bluestar, Dow Corning, Evonik Industries, Momentive, Shin-Etsu and Wacker.

2.1.1 Methyl-silicone resins

Methyl-silicone resins are the polymethylsiloxanes with the lowest carbon content. Their long-term heat resistance is between 180 to 200 °C. Exposure to higher temperatures results eventually in complete oxidation of the methyl groups[20].

The chemical analogy to silica determines the partially inorganic character of this group of resins. This explains properties such as the relatively high hardness, low thermoplasticity, poor pigmentability and incompatibility with organic resins. The methyl-silicone resins exhibit excellent water repellency even in a solvent-free, partially crosslinked state at room temperature.

In principle, methyl-silicone resins can be subdivided into two types: ambient temperature curing systems and systems curing at high temperatures (baking systems). The decisive structural differences between the two types are, as shown schematically in Figure 2.5, the functionality density and the molecular weight of the silicone resins.

The ambient curing systems are high alkoxy-functional low-molecular silicone resin building blocks whose chemical composition corresponds to that of the silicone resin intermediates described in Chapter 1.3. Consequently with these products, a pronounced build-up of molecular weight, as occurs with "bodying" is dispensed with. The alkoxy content of ambient curing methyl-silicone resins is

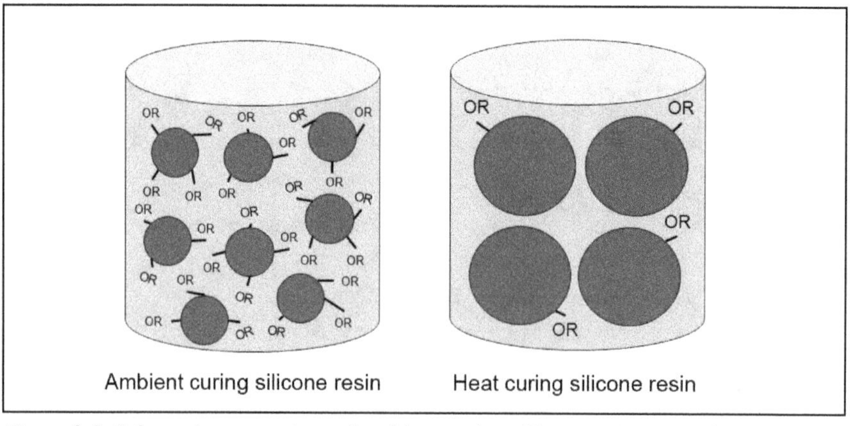

Figure 2.5: Schematic comparison of ambient curing silicone resin versus heat curing system

generally approximately 30 % w/w. Currently there is substantial industrial use of methoxy-functional ambient curing methyl-silicone resins but ethoxy-functional derivatives are of increasing interest. Suitable catalysts, such as tetra n-butyl titanate (TnBT) are used to accelerate curing in the presence of atmospheric moisture (see Figure 2.7). Curing cannot take place without atmospheric moisture as water is required to hydrolyze the alkoxy groups of ambient curing silicone resins and condensation of the silanol groups formed only occurs subsequently during curing of the film. Ambient curing methyl-silicone resins are commercially available with a solids content of up to 100 % w/w. The very low viscosities despite high active substance content are explained by the low molecular weights.

Compared with ambient temperature curing resins, methyl-silicone resins, which are oven-cured at high temperatures, have markedly higher molecular weights and only a very low functional density of alkoxy or silanol groups. Typically these resins are baked for 30 minutes at approximately 250 °C to obtain a hard, chemically resistant coating. The high temperature curing methyl-silicone resins are commercially available mainly in aromatic or aliphatic solvents or alcohols.

The different curing conditions and processes are shown schematically in Figure 2.6. Physical drying should be understood as evaporation of solvent from the coating.

The advantages resulting from drying at room temperature are obvious. Oven-drying at high temperatures limits the size of the item to be coated to the dimensions of the available oven. In the ambient temperature curing process, large items can be coated with silicone resins in the open air. The use of ambient curing

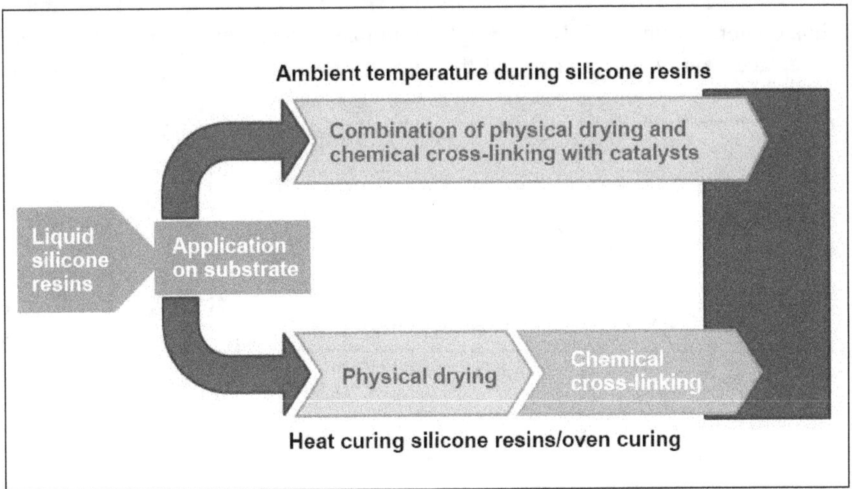

Figure 2.6: Drying mechanism of ambient curing silicone resins versus heat curing silicone resins

resins instead of oven curing at high temperatures can markedly reduce energy consumption.

Substrate pre-treatment is crucial for methyl-silicone resins. Substrates must be shot-blasted. In general methyl-silicone resins are very sensitive to oil, grease or welding residues on the substrate. Phosphating pre-treatments are not suitable for high temperature applications as they lead to adhesion problems.

For applications involving extended exposure to temperatures above 200 °C, the use of at least 15 % w/w (solid/solid) laminar pigments/fillers, such as mica, is necessary. The pigment volume concentration of the coating should be between 20 to 30 % w/w. As a general rule the dry film thickness should be 25 ± 5 µm.

In thicker films, methyl-silicone resins tend to form stress cracks and exhibit increased notch sensitivity especially under sudden thermal stress.

2.1.2 Methyl/phenyl-silicone resins

To be classified in this group of resins, the phenyl group content is generally above 20 %[20]. The other organic groups are mainly methyl. Silicone resins which only contain phenyl groups have found only very limited applications as their particular disadvantage is long lasting thermoplasticity. The phenyl groups in the methyl/phenyl-silicone resins raise the heat resistance to 200 to 250 °C and significantly improve compatibility with other organic polymers. Due to the improved compatibility with other organic resins methyl/phenyl-silicone resins are frequently used for the synthesis of silicone combination resins. Methyl/phenyl-resins exhibit a solids content of up to 90 % by weight. Suitable solvents are primarily aromatics, replacement of which is one of the important tasks in developing new, more

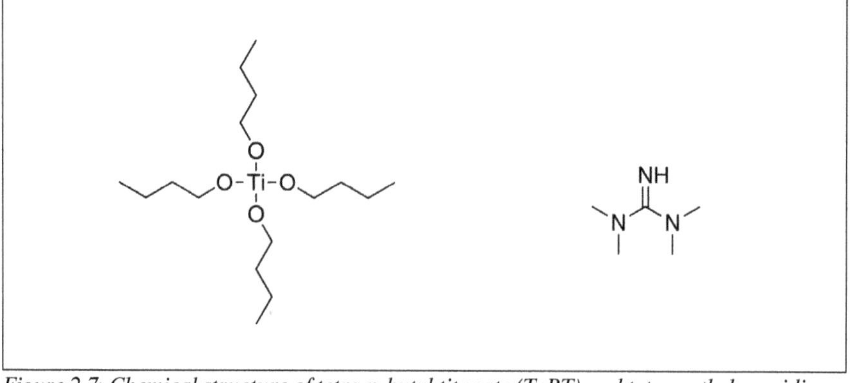

Figure 2.7: Chemical structure of tetra n-butyl titanate (TnBT) and tetramethylguanidine (TMG)

environmentally-friendly products. Generally methyl/phenyl-silicone resins solutions are not miscible with methyl-silicone resin solutions. Incorporation of difunctional monomers (D-units) in the methyl/phenyl-silicone resins enables the properties of the resin films to be adjusted from hard to very elastic.

Most of the commercially available methyl/phenyl-silicone resins must be oven-dried at high temperatures (30 min, 250 °C). Only with the latest developments has it become possible to cure methyl/phenyl-silicone resins at ambient temperatures on an industrial scale. Mixtures of tetra n-butyl titanate (TnBT) and tetramethylguanidine (TMG) have proved particularly suitable catalysts (Figure 2.7).

With this combination the TnBT reacts as a Lewis-Acid by forming chemical bonds with the polymer and the TMG reacts as a strong base accelerating the reaction. Both catalysts are soluble and miscible in xylene.

The structural differences of the ambient curing/oven curing methyl/phenyl-silicone resins correspond to those shown for methyl-silicone resins in Figure 2.5.

2.2 Silicone combination resins/ silicone resin hybrids

A very large number of products is described under this term in the literature and used in industry. In principle, there is an almost unimaginably vast range of possibilities since the molecular size, chemical structure and functionality of the possible educts (polysiloxanes and organic polymers) can be greatly varied. Potential silanol-functional polysiloxanes include silicone resin intermediates or linear polysiloxanes. The polysiloxane content of silicone combination resins can range from less than 5 % to more than 90 %.

The basic idea of silicone combination resins is to unite the advantages of polysiloxane chemistry with those of organic resins. For certain silicone combination resins and their coating applications this can lead to outstanding performance.

The polysiloxane content in silicone combination resins improves, for example, the heat stability, weathering resistance, water repellency, release and easy-to-clean properties of coatings. A further increasingly important aspect is the reduction of VOCs, achieved by the polysiloxane presence in silicone combination resins. With silicone combination resins, very high solids content in the binder and final coating formulation are often possible at low viscosities, a situation impossible to achieve with most of pure organic binders. This is because the viscosity is reduced by the polysiloxane component in silicone combination resins. In view of the increasingly stringent regulations concerning reduction of VOCs, this aspect will further reinforce the interest shown in recent years in silicone combination resins.

Suitable choice of organic resins can improve drying conditions, compatibility with organic resins, substrate adhesion, pigmentability, solvent resistance and, not least, "attenuate" the higher prices of 100 % silicone resins.

Another crucial, much discussed point is whether the organic resins (e.g. alkyd resins, polyesters or epoxy resins) must react chemically with a suitable polysiloxane or whether a simple cold blend of a polysiloxane with an organic resin is adequate[21]. In the case of a cold blend, chemical reaction of the polysiloxane with the organic resin takes place during curing of the coating.

The most decisive argument for chemical coupling of the polysiloxane with the organic resin is the much improved compatibility. In most cases a cold mixture is very unstable and separates quickly. This is because polysiloxanes essentially have very limited compatibility with organic compounds, irrespective of whether they are hydrophilic or hydrophobic. Certain mixtures can be maintained in a stable state for an adequate interval through correct choice of solvents. There is a danger, however, that during the drying process, separation of the silicone resins and the organic resins will occur and that domains of silicone resins and organic resins will form in the cured film.

For example, it was found that the thermal stability in terms of yellowing of cold blends of polyester and silicone resin intermediates was insufficient. A possible cause is the formation of micelles from the resins which are present next to each other in unreacted form and which can be regarded as individual resins as far as their chemical behavior is concerned.

In general, it can be postulated that incompatible resin mixtures can only combine their individual advantages when both partners react together chemically.

The silanol-functional polysiloxanes (silicone resin intermediates or linear polysiloxanes) can be modified with numerous organic C-OH functional resins resulting, under specific conditions, in stable resin end products. The use of three dimensional silicone resin intermediates in coatings is often preferred to that of linear polysiloxanes. This is because the linear structures produce relatively soft films whereas the three dimensional networks of the silicone resin intermediates permit hard, durable coatings.

The oligosiloxanols are modified chemically with organic resins according to the reactions described in Figure 2.8.

As shown in Figure 2.8, homocondensation of the silanol groups of the oligosilanol takes place as a competing reaction alongside the condensation reaction of the organic C-OH functional resins[22].

Choosing the right reaction parameters allows competitive homocondensation of silanol groups to be minimized as much as possible during modification. Mutual

$$\equiv Si-OH + HO-C\equiv \longrightarrow \equiv Si-O-C\equiv + H_2O$$

$$\equiv Si-OH + HO-Si\equiv \longrightarrow \equiv Si-O-Si\equiv + H_2O$$

Figure 2.8: Condensation reaction between organo-siloxanols and an OH-functional organic resin and homocondensation of silanol groups

condensation of the organic polyols, i.e. ether formation, is not generally observed under the conditions in which modification is carried out.

The silicone combination resin synthesis is problematical with regards to both obtaining a stable reproducible final resin state and the degree of crosslinking in the cured combination resin. Although interruption of the modification reaction is achieved by rapid cooling, it is nevertheless difficult to bring this about reproducibly with always the same degree of crosslinking. From a chemical viewpoint, it is obvious that the reaction of polysilanol-functional polysiloxanes with polyfunctional organic resins can easily lead to an uncontrolled increase in molecular weight or to gelling of the product.

The non-reproducible competition between the modification reaction and silanol condensation results in non-uniform resin end products, a problem which continues during curing. As with the pure silicone resins, residual silanol groups in the cured film can further condense, especially under thermal stress, leading to embrittlement.

These difficulties can be minimized to some extent by the use of alkoxy-functional organo-siloxanes as a silicone intermediate (see also Chapter 1.3). Modification and curing also proceed by the same chemical reaction when this route is followed:

$$\equiv Si-OR + HO-C\equiv \longrightarrow \equiv Si-O-C\equiv + ROH$$

Figure 2.9: Reaction between alkoxy-functional polysiloxane and an OH-functional organic resin

Silicone combination resins with polyester resins, epoxy resins and acrylic resins have generated the greatest interest commercially.

2.2.1 Silicone-modified polyester resins

Numerous silicone-modified polyester resins are described in the literature and used commercially. This is because of their diverse potential applications and the good property profiles which can stem from their use. With these resins the advantages of silicones such as high temperature resistance, excellent weathering resistance and low surface tension can be combined with the advantages of polyester resins such as low thermoplasticity, high flexibility and good pigment wetting. Cold blends of organic polyester resins and silanol/alkoxy-functional polysiloxanes as well as chemically linked reaction products of silanol/alkoxy-functional polysiloxanes with organic polyester resins are used.

The choice of polyester resin decisively determines the property profile of the resultant polymers. This class of binders can be divided into two groups:

1. Combinations with aromatic polyester resins
2. Combinations with aliphatic polyester resins

In the first case the silicone-modified aromatic polyester resins are synthesized in a two-step procedure. In the first step, suitable alkoxy-functional silicone intermediates of a defined molecular weight are produced. Generally methoxy-functional methyl/phenyl-resin intermediates are used.

In the second step, these intermediates are reacted with OH-functional aromatic polyesters based on aromatic dicarboxylic acids (e.g. isophthalic and terephthalic acid). Polar solvents, including small quantities of alcohols, are used to prevent undesirable side reactions and to maintain viscosity stability during resin manufacture.

The chemical reaction is shown schematically in Figure 2.10 and is described in detail in the patent literature[23].

The molar mass is strongly increased by the modification reaction. Since the final combination resins contain alkoxy and hydroxyl groups, they can be thermally self-cured. In addition, crosslinkers, such as melamine resins, can be used to adjust the final coating properties via the crosslinking density.

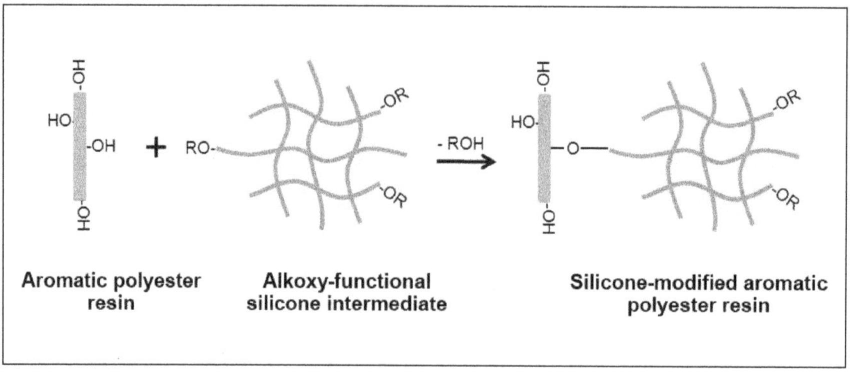

Figure 2.10: Synthesis of a silicone-modified aromatic polyester resin

Silicone-modified aromatic polyester resins are frequently used for coatings with a continuous heat resistance up to 250 °C, depending on the silicone content which can vary between 10 and 80 % w/w. Heat resistance improves with increasing silicone content.

A further characteristic of coatings based on silicone-modified aromatic polyester resins is their low thermoplasticity. As both silicone and polyester portions are strongly crosslinked, hardness is maintained even at temperatures around 150 °C. This is important in applications in which the hot coatings are mechanically stressed but must be scratch-resistant. The properties of silicone-modified aromatic polyester resins are particularly advantageous in decorative coatings of thermally stressed appliances such as toasters, fan heaters and cookers as well as the outer coatings of deep fryers, pots and pans.

The use of silicone-modified aromatic polyester resins in industrial top-coats is limited by the inadequate weather resistance of the applied coatings.

This technology is commercially available from, for example, Eternal, Evonik Industries and Shin-Etsu.

In the second case, advantage is taken of the highly attractive combination of the weathering resistances of aliphatic polyester resins and silicone resins.

Silicone-resin modified aliphatic polyester resins can be produced by a synthesis analogous to that shown in Figure 2.10 for silicone-modified aromatic polyester resins. These compounds are hydroxy- and alkoxy-functional and are cured via heat drying as described previously for aromatic derivatives.

Another very interesting class of compounds, the silicone-modified aliphatic polyester resins shown schematically in Figure 2.11, do not have any alkoxy-functionalities in the silicone resin part but only hydroxy groups on the polyester part.

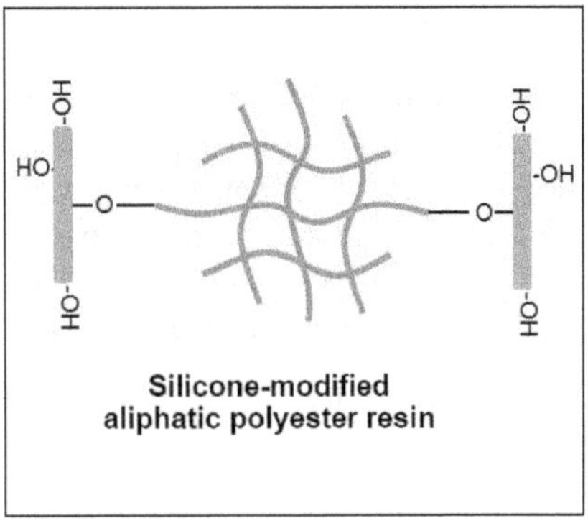

Figure 2.11: Schematic structure of a silicone-modified aliphatic polyester resin

The hydroxy groups can be crosslinked, for example with isocyanates at room temperature. Consequently, these silicone-modified polyester resins can be used like classic polyester polyols but have the advantage of bringing the positive effects of silicone resins. Furthermore very high solids content (>85 %) at acceptable viscosities can be achieved with this class of binders. This aspect is likely to be increasingly important over the next few years because of legislation regarding reduction of VOC levels. Another advantage is the broad compatibility of silicone-modified aliphatic polyester resins which facilitates their use as a co-binder in a wide range of coating formulations.

Typical fields of application of these compounds are top-coats such as anti-corrosion coatings, marine coatings, agricultural equipment coatings, transportation and wood coatings.

These raw materials are available commercially from, for example, Evonik Industries and Synthopol.

2.2.2 Silicone-modified alkyd resins

Silicone-modified alkyd resins are manufactured by reacting medium- (40 to 60 % oil) and long-oil (over 60 % oil) alkyd resins, based on, for example, soya bean oil, safflower seed oil or sunflower oil, with silanol-functional polysiloxanes at 160 to 180 °C. In general methyl/phenyl-silicone resin intermediates are used as polysiloxane components. The hydroxyl number of the used alkyd resins must be higher than that of the silanol methyl/phenyl-silicone resin intermediates due to gelling. Low molecular weight silicone resin intermediates must be used since the condensation results in considerable molecular enlargement. The alkyd resin/polysiloxane mixing ratio is approx. 70:30.

The alkyd resin and silicone resin educts are initially mutually incompatible and form a homogeneous clear resin only after partial linking. To obtain resins free of cloudiness, the solubility parameters of the combination partners should not differ greatly[24].

Modification of alkyd resins with silicone resin intermediates leads to improved touch-drying times, resistance to moisture, chemicals and weathering. However, the presence of other constituents means that improvements in heat resistance are less than, for example, with silicone-modified polyester resins[25]. The use of this resin class is limited in industry. Their main applications are in decorative and architectural paints, marine paints and two-pack wood finishes[26].

These products are available commercially from, for example, Arkema and Worlée Chemie.

2.2.3 Silicone-modified epoxy resins

Numerous silicone-modified epoxy resins are described in the literature and used commercially. This is because of the often outstanding properties which can be achieved with this class of products. Cold blends of organic resins and silanol/

alkoxy-functional polysiloxanes as well as chemically linked reaction products of silanol/alkoxy-functional polysiloxanes are used.

The choice of epoxy resin decisively determines the property profile of the resultant polymers. This class of binders can be divided into two groups:

1. Combinations with aromatic epoxy resins
2. Combinations with aliphatic epoxy resins

In the first case, the molar mass is strongly increased by the hydrolysis/condensation reaction of alkoxy-functional silicone resin intermediates with hydroxy-functional aromatic epoxy-resins. The resulting resins are thus highly viscous and exhibit outstanding physical surface-drying characteristics.

The amount of silicone in this type of resin can vary between 30 and 65 % w/w. The main area of application of this heat resistant silicone combination resin is in high temperature anti-corrosion coatings. Additional properties obtained by the combination of silicone resins with epoxy resins and which clearly differentiate this class of resins from pure silicones are very good adhesion, flexibility and easy pigmentability. Such resins can be cured not only thermally but also with amines at room temperature. However the use of the silicone aromatic epoxy resins is limited due to insufficient weather resistance and yellowing. These disadvantages significantly limit the potential use in industrial top-coat applications.

Such products are commercially available from, for example, Evonik Industries and Shin-Etsu.

In the second case, the previously mentioned disadvantages of silicone-modified aromatic epoxy resins can be overcome by using silicone-modified aliphatic epoxy resins based on hydrogenated bisphenol and alkoxy-functional silicone resin[27–29].

The first step in the manufacture of chemically-linked silicone-modified aliphatic epoxy resins is the synthesis of alkoxy-functional silicone resin intermediates of a defined molecular weight. In a second step, these resins are reacted with OH-functional aliphatic epoxides to form the silicone epoxy hybrid resin as shown schematically in Figure 2.12[30]. As already mentioned in Chapter 2.2, the molecular structure and optional gelling must be kept in mind when managing the reaction.

The silicone-epoxy resin consequently consists of a chemically-linked silicone resin and aliphatic epoxy resin components which contain epoxy, alkoxy and hydroxy groups for curing (see Figure 2.1.2).

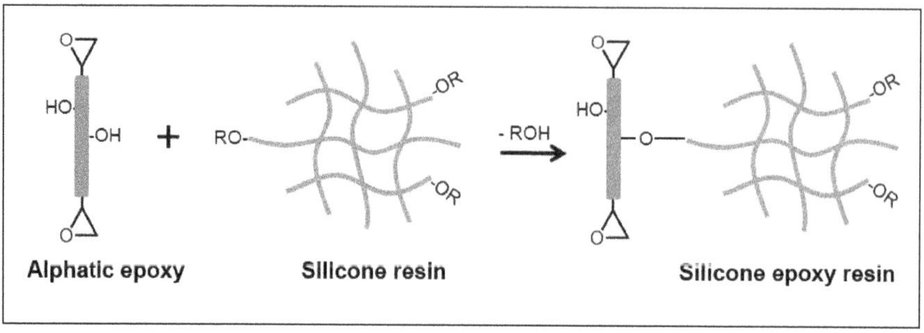

Figure 2.12: Synthesis of a silicone-modified epoxy hybrid

These silicone-epoxy resins have various advantages over physical mixtures (cold blends) of alkoxy-functional silicone resins and epoxy-functional compounds:

- No phase separation between the organic and silicone phases is observed.
- No micelle structure from parallel organic and silicone domains which cause inadequate chemical and weathering resistance.

The hardeners used for silicone-epoxy hybrids are generally amino-trialkoxy silanes.

The amino group of the amino-trialkoxy silane reacts with the epoxy group of the silicone-modified epoxy resin as shown schematically in Figure 2.13. The alkoxy groups of the amino-trialkoxy silane then react by hydrolyzing/condensing with the free alkoxy groups of the polysiloxane in the presence of atmospheric moisture. The amine/epoxy addition reaction occurs simultaneously at room temperature with the hydrolysis/condensation reaction of the alkoxy groups from the amino-silanes and silicone hybrid resins. The hydrolysis/condensation reaction can be catalyzed by Lewis acids, acids, or amines and drying is consequently accelerated.

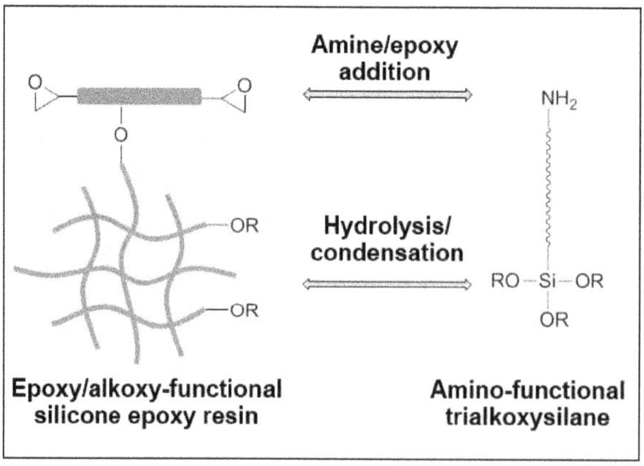

Figure 2.13: Dual curing mechanism of silicone-modified epoxy resins with amino-functional trialkoxysilanes at room temperature

This “double crosslinking” results in a silicone-epoxy framework with an extremely high degree of crosslinking. Also it enables hard coatings with high abrasion resistance and chemical resistance to be achieved.

The aliphatic epoxy components in the silicone-epoxy hybrid impart good corrosion and chemical resistance, high pigmentability and wettability, good mechanical characteristics as well as outstanding adhesion to metal or the primer coat.

The silicone resin building blocks and the silane curing agent enable low formulation viscosities to be achieved which makes possible very low VOC content for ultra-high solid coating formulations. The dried coatings based on this binder show outstanding anti-corrosion properties, good UV resistance, low yellowing, high heat resistance, low flammability and stain resistance[21]. The amino-functional trialkoxysilane hardener improves the adhesion and resistance to corrosion and chemicals.

Therefore, the strong features of three different technologies are united in one system and generate the excellent properties of this resin class.

Analogous cold blends of alkoxy-functional silicone resins, aliphatic epoxy resins and amino-functional alkoxysilanes are described in the literature[31, 32] and used frequently in industry.

The same crosslinking reactions occur with these cold blends, as those shown schematically in Figure 2.13. However great care must be taken that the components which are only physically mixed do not separate in the formulation or during crosslinkage of the coatings as, otherwise, surface properties will be inadequate.

Products based on this technology are available, for example, from Evonik Industries and PPG Industries.

A very interesting alternative to the combination of organic epoxy resins and siloxane resins described before is to use epoxy-functional silanes (e.g. glycidoxypropyl trialkoxysilane) at the silicone resin synthesis stage. This permits epoxy-functional silicone resins to be synthesized. This technology (Figure 2.14) is frequently described in the patent literature[33, 34]. Amino-functional agents, such as amino-functional trialkoxysilanes, are also used in this method of curing the coatings. A further modification of this technology is the combination of epoxy-functional silicone resins and amino-functional hardeners with organic acrylic resins[34]. Waterborne emulsions of epoxy-functional silicone resins are also described in the patent literature[35].

This type of product is available from, for example, Dow Corning and Wacker.

Another technique frequently used for silicone-modified epoxy resins is cold blending of organic epoxy resins with amino-functional silicone resins. The amino-functional methyl/phenyl-silicone resins used are obtained by reacting silanol-functional silicone resin intermediates with amino-functional trialkoxy-silanes[36]. In this method the amino-functional silicone resin acts as a hardener, the silicone resin nature of which improves the binder properties of organic epoxy resins. This technology results in high solids coatings with excellent weathering resistance and anti-corrosion properties.

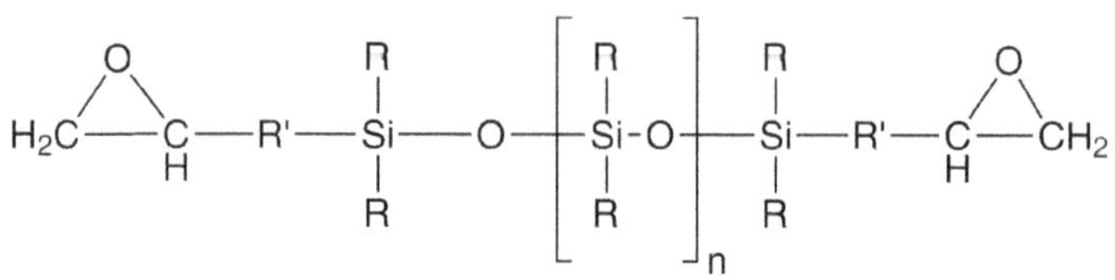

R = Alkyl, aryl, vinyl, glycole, alkoxy ($C_1 - C_8$) or epoxy
R' = Alkene ($C_1 - C_{18}$), optionally interspersed by oxygen or arylene groups
n > 1

Figure 2.14: Structure of epoxy-functional silicone resin

These raw materials are available from, for example, Dow Corning and Wacker.

The silicone-modified aliphatic epoxy resins are typically used in a number of industrial premium top coat applications such as high performance anti-corrosion coatings, anti-graffiti and foul release coatings (due to the easy-to-clean properties), concrete-, floor- and marine-coatings.

2.2.4 Silicone-modified polyacrylate resins

Numerous silicone-modified polyacrylate resins are used commercially. This is because of the diversity of potential applications and the good property profiles which can stem from their use.

With silicone-modified polyacrylate resins, the advantages of silicones, such as high heat resistance, excellent weathering resistance, low surface tension, can be combined with the advantages of polyacrylate resins, such as fast complete drying, good weathering resistance, high stability to hydrolysis, excellent hardness and good pigment wetting.

Cold blends of polyacrylate resins and silanol/alkoxy-functional polysiloxanes as well as chemically linked reaction products of silanol/alkoxy-functional polysiloxanes with polyacrylate resins are used[37]. Both "pure" silicone-modified polyacrylate resins as well as silicone combination resins which contain polyacrylate and polyester building blocks are described in the literature[38, 39].

Such raw materials are commercially available from, for example, from Evonik Industries, Synthopol and Wacker.

Another frequently used method is the cold blending of alkoxy-functional silicone resins with oligoacrylic-functional organic building blocks (e.g. tripropyleneglycol diacrylate) and amino-functional trialkoxysilanes[40, 41]. The amino group of the amino-functional trialkoxysilanes reacts with the acrylic function of the oligoacrylic-functional organic building block by a Michael-type addition reaction. The alkoxy groups of the amino-trialkoxy silane then react by hydrolyzing/condensing in the presence of atmospheric moisture with the free alkoxy groups of the silicone resin. Both reactions occur simultaneously at room temperature. The hydrolysis/condensation can be accelerated by, for example, titanium or tin catalysts. This "double crosslinking" results in a high degree of crosslinking and produces hard coatings with high abrasion resistance and chemical resistance. Such coatings are frequently used to protect steel structures and/or as weather-resistant exterior coatings.

2.3 Alkoxy-silyl modified resins and alkoxy-silyl modified isocyanate crosslinkers

Various alkoxy-silyl modified resins and alkoxy-silyl modified isocyanate crosslinkers are described in the literature and used industrially. These are not classic silicone resins, such as those described in Chapters 2.1 and 2.2, but organic resins or isocyanate crosslinkers, which have been modified with alkoxy-functional silyl groups. Consequently, no polysiloxane structures but single alkoxy-functional silyl groups bound to organic molecules are present. The aim of alkoxy-silyl modification is to incorporate additional inorganic properties and further curing mechanisms (hydrolysis/condensation or transesterfication) in standard polyurethane coatings systems and to form interpenetrating networks.

The most commercially significant application for such coatings systems is in modern automotive top coats. This is because the colour and gloss of an automotive finish have a decisive influence on customers' choice of car[42]. However, the appearance of an automotive finish deteriorates with time as it is exposed to numerous external influences. These include sun, rain, snow, and others such as tree resin, bird droppings and salt which all attack the paint. Another significant problem is the very small scratches caused by dirt on the surface of the car body as soon as, for example car-wash brushes clean the car. If the scratches mount up over time, the automotive finish looks dull and shabby.

The polyurethane-based clear coat, the topmost layer of an automotive finish, should protect against mechanical and weather-related attack.

Traditionally used polyurethane-based clear coats are well known for their good chemical resistance, weathering resistance and physical properties. Scratch resistance can be improved by the use of innovative alkoxy-silyl modified resins or alkoxy-silyl modified isocyanate crosslinkers in polyurethane coatings instead of standard polyurethane systems. The organic part of such hybrid materials is between ca. 90 to 95 %. The inorganic silicon-based part is about 5 to 10 %[42].

Various technologies are used and described in the literature. The most important are:

A) Modification of poly(meth)acrylate resins with alkoxy-silyl groups and their use in automotive clear coats is described in the patent literature[43]. In this approach (meth)acrylic-functional alkoxy silane monomers are used in the synthesis of polyacrylate resin for subsequent crosslinking with polyisocyanates[44, 45]. However, because of the broad molecular weight distribution of alkoxy-silyl modified poly(meth)acrylate resins, only low solids content is possible. At higher solids content, the coatings are difficult to work with because of high viscosities. [46].

Alkoxy-silyl modified polyacrylate resins can also be used for weather and stain resistant construction and heavy duty coatings.
Such alkoxy-silyl modified polyacrylate resins are commercially available from, for example, Kaneka.

B) The latest generation of alkoxy-silyl modified resins consists of reaction products of oligoalcohol molecules with isocyanate-functional tri-alkoxy silanes [47, 48] (see Figure 2.15). This results in alkoxy-silyl-functional urethane resins which, as

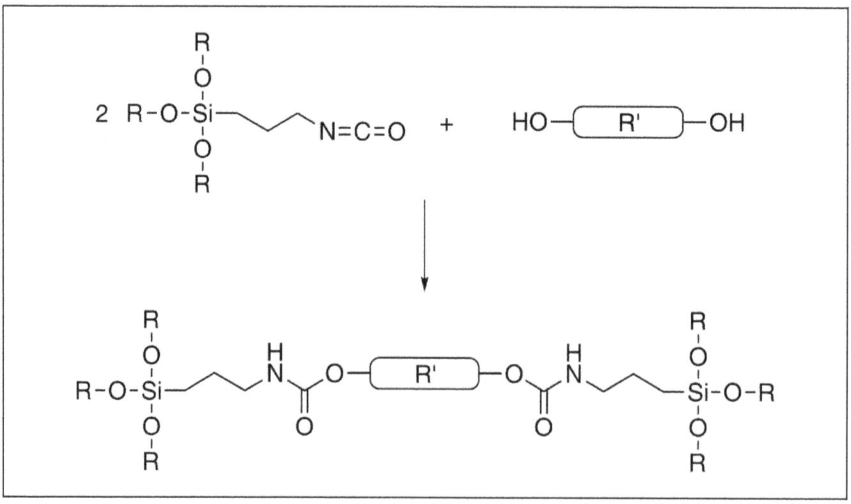

Figure 2.15: Synthesis of alkoxy-silyl-functional urethane resins

co-crosslinkers can be used alongside an isocyanate crosslinker, in combination with a suitable catalyst and polyol resins, in automotive clear coat formulations. These alkoxy-silyl-functional urethane resins permit the manufacture of automotive clear coats with higher scratch resistance, chemical resistance and good adhesion[47].

The alkoxy-silyl-functional urethane resins can also be used in wood coatings, plastic coatings and high-tech adhesives. One reason for their versatility is the various curing processes that can be used with the alkoxy-silyl-functional urethane resins. Besides the use as a boosting co-crosslinker in combination with an isocyanate crosslinker, a polyol and a catalyst for two-pack polyurethane systems described above, alkoxy-silyl-functional urethane resins can also be used in non-isocyanate curing systems. Drying at room temperature is also possible with alkoxy-silyl-functional urethane resins.

Such alkoxy-silyl-functional urethane resins are commercially available from, for example, Evonik Industries.

C) Alkoxy-silyl modified isocyanate crosslinkers are produced by reacting, for example, trimerized hexamethylene diisocyanate (HDI) or isophorone-diisocyanate (IPDI) with alkoxysilane-functional secondary amines (e.g. bis[3-(trimethoxysilyl)propyl]amine)[49–51]. The secondary amines react with the isocyanate groups to form urea groups. As a rule, the reaction conditions are chosen so that the isocyanate groups react only partially with the secondary amines. The resultant alkoxy-silyl modified isocyanate crosslinkers can subsequently be cured with polyols and suitable catalysts.

Such alkoxy-silyl modified isocyanate crosslinkers are commercially available from, for example, Evonik Industries.

2.4 Radiation-curable silicone resins

2.4.1 Acrylic-functional silicone resins

Acrylic-functional silicone resins are reactive resins based mainly on linear polysiloxane structures. Over the last two decades they have attracted increasing attention in industry. Curing takes place by free radical polymerization using photocatalysts.

Due to their good release properties, acrylic-functional silicone resins are frequently used in release coatings for numerous self-adhesive products.

The key advantage of UV-curable acrylic-functional silicone resins is that they can be cured solvent-free at room temperature. Oxygen interferes with the reaction, so an inert nitrogen environment is required.

For release liners cold cure permits the use of heat sensitive substrates such as PE, PP or PVC films. Paper substrates do not need to be re-moisturized and show absolute lay-flat characteristics. The very fast cure allows the use of acrylic-functional silicone resins in inline processes together with adhesive coating. It also makes very high line speeds possible. Typical fields of application are release liners for self-adhesive labels, graphic arts, tapes and hygienic products.

Another very important use of the acrylic-functional silicone resins are crosslinkable wetting, flow and slip additives for radiation cured coatings and printing inks. Without additives[52], solvent-free UV-systems would be inadequate for most applications due to their insufficient substrate wetting and flow characteristics[53, 54].

First generation acrylic-functional silicone resins were based on a condensation reaction between linear chloropolysiloxanes and OH-containing acrylates, for example 2-hydroxyethylacrylate or pentaerythritol-triacrylate (see Figure 2.16). However these structures suffered the disadvantage of inadequate hydrolysis stability due to the hydrolysis sensitive SiOC bonds[55, 56].

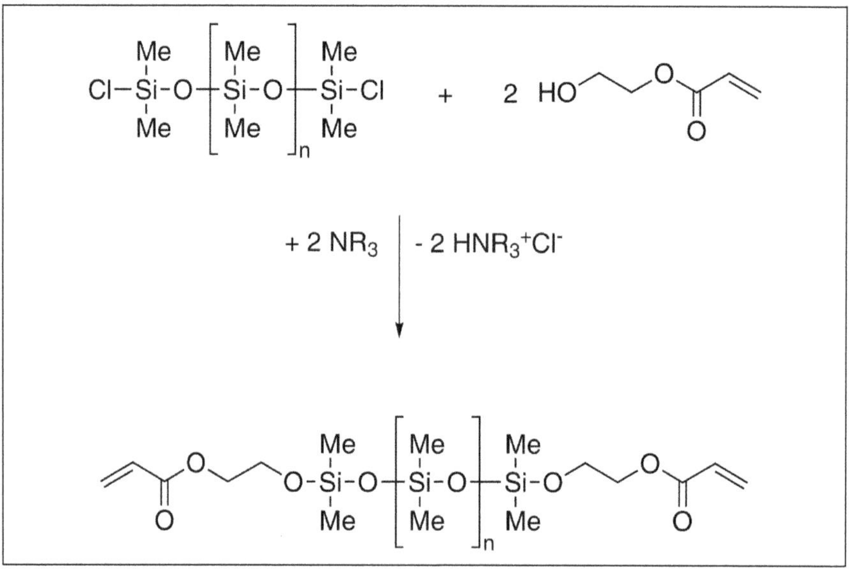

Figure 2.16: First generation acrylic-functional silicone resins

Later generations are synthesized via SiC chemistry. Resulting products are significantly more resistant to hydrolysis than the first generation. In principle, SiC chemistry also offers a wider possible range of structural variation than SiOC chemistry thus permitting the chemical structure of acrylic functional resins to be tailored more closely to the desired application properties.

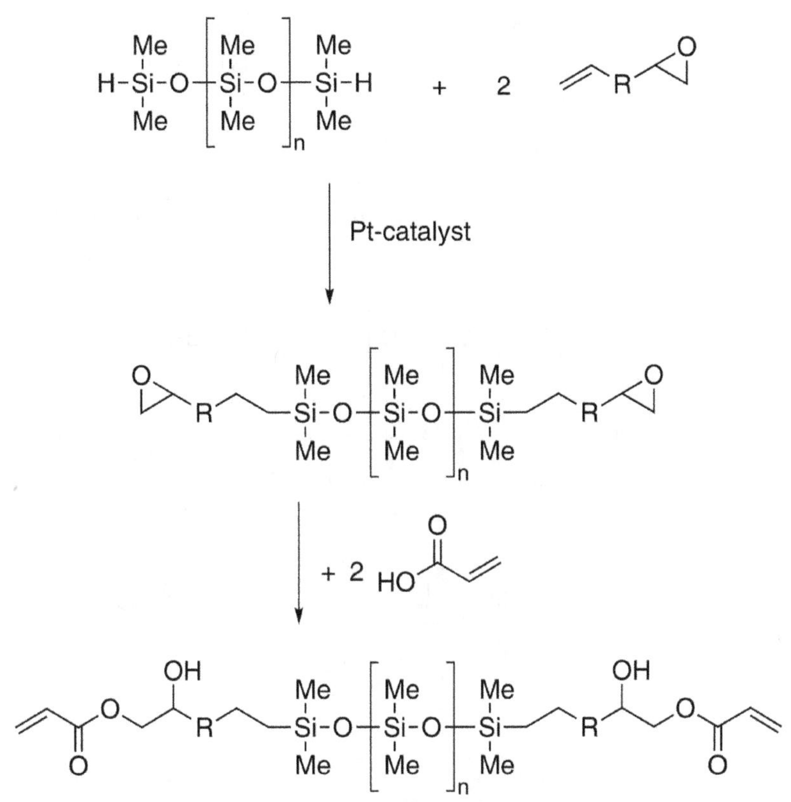

Figure 2.17: Second generation acrylic-functional silicone resins

Figure 2.17 shows schematically the manufacture of typical acrylic-functional silicone resins via a hydrosilylation reaction of an unsaturated epoxide and a linear SiH-functional polysiloxane, followed by catalytic opening of the epoxide group by acrylic acid with the formation of an ester. An alternative to the previously described synthesis path is to esterify with acrylic acid hydroxy-functional polysiloxanes. The latter are manufactured by reacting unsaturated alcohols (such as allyl alcohol or 1-hexenol) with SiH-functional polysiloxanes via hydrosilylation. Because of its wide use in industry, the manufacture of acrylic-functional silicone resins via SiC chemistry is comprehensively documented in the patent literature[57–59].

The latest product generation of acrylic-functional silicone resins has multi-acrylated centres which are coupled to a linear polysiloxane via SiC chemistry

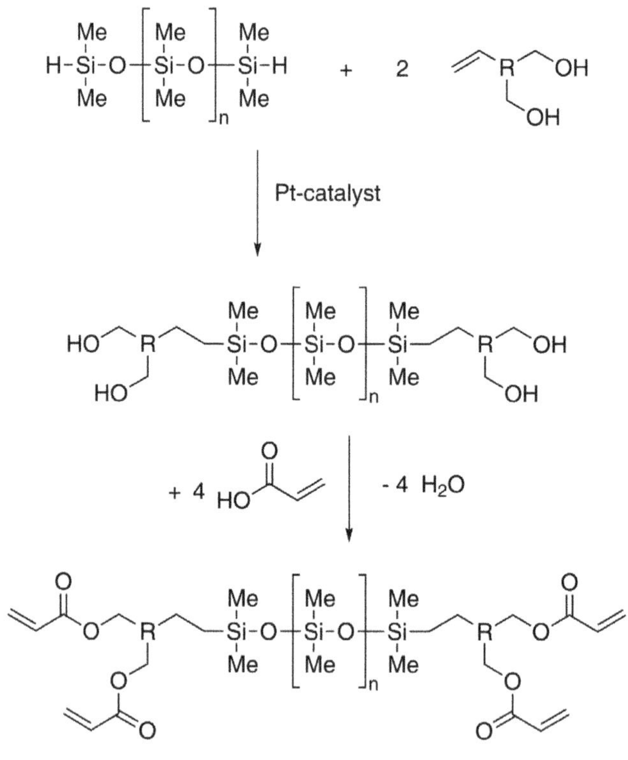

Figure 2.18: Latest generation acrylic-functional silicone resins

(see Figure 2.18). These structures show significantly improved end properties such as good aging, premium release and very fast and effective curing in comparison with those of the mono-acrylic products described above[60, 61].

In some specific release liner applications a higher release value (controlled release) is required. This effect can be generated by acrylic-functional silicone resins synthesized using three dimensional silicone resin intermediates as one of the starting materials. The synthesis of these structures is also described in the patent literature[62, 63].

Currently, acrylic-functional silicone resins cover the whole range of release levels (premium/easy/controlled/tight release applications). There are complete systems based on various acrylic-functional silicone resin components which can be mixed with each other to tailor the release characteristics to specific applications. Fully-cured and co-polymerized coatings are characterized by good mechanical processing and applicational properties.

Acrylic-functional silicone resins are commercially available from, for example, Bluestar Silicones, Evonik Industries and Shin-Etsu.

2.4.2 Epoxy-functional silicone resins

Epoxy-functional silicone resins are, like the acrylic-functional silicone resins, described in the previous section, reactive resins. As a rule, they are based on linear polysiloxane structures. They can be radiation-cured in the presence of cationic photocatalysts.

These products are also used as coating compounds for release coatings for numerous self-adhesive products.

A complete system of different structures and photocatalysts covering a broad range of release values (from premium to tight release) is commercially available for cationic-curing epoxy-functional silicone resins. Nearly all required release values can be obtained by simply mixing commercially available materials. Substrates, especially papers, must be checked for impurities that might affect curing. Oxygen does not interfere with the reaction, so there is no need for inerting with nitrogen.

The manufacture of typical epoxy-functional silicone resins via hydrosilylation of vinyl cyclohexane monoxide (VCMO) is shown schematically in Figure 2.19. In contrast to the acrylic-functional silicone resins, this is a one-step reaction.

The manufacture of epoxy-functional silicone resins and their use as release liners and coatings additives are described comprehensively in the patent literature[64].

The epoxy-functional silicone resins are commercially available from, for example, Bluestar Silicones, Evonik Industries and Momentive.

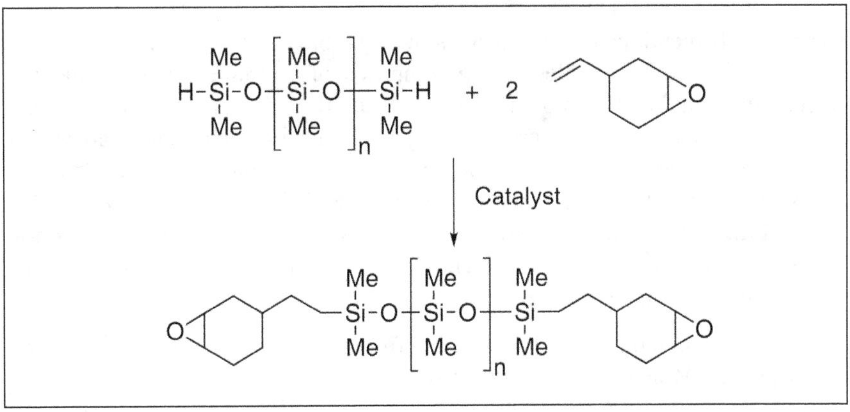

Figure 2.19: Synthesis of epoxy-functional silicone resins via hydrosilylation of VCMO

2.5 Room temperature vulcanizing silicone resins (RTV resins)

The RTV (room temperature vulcanizing) silicone resins[65–67] are well-known in the fields of casting compounds and the manufacture of sealants. However, specific properties of these polymers are sometimes useful in coating formulations and are frequently used, for example, for release liners.

The cold vulcanizing two-component silicone rubbers are castable, brushable, kneadable compounds. After adding the curing component, they react to form an elastic silicone rubber. With addition crosslinked types, accelerated vulcanization at raised temperatures is frequently used.

The most important properties of elastic silicone rubber are

- permanent flexibility
- electrical insulation
- release effect
- adhesion to substrate
- chemical resistance.

There are several types of crosslinking[68]:

1) addition crosslinking (between vinyl groups of the polysiloxane and Si-H-functional crosslinker)
2) crosslinking with peroxides (between vinyl groups of the polysiloxanes)
3) condensation crosslinking (with α, ω-dihydroxy polysiloxanes and silica esters)

Addition crosslinking and condensation crosslinking require a suitable catalyst in addition to the corresponding educts and reaction conditions. In the former, this can be a platinum catalyst, in the latter a tin catalyst.

As previously mentioned, addition crosslinking is based on linking Si-H groups to vinyl double bonds[69]. Besides salts or complexes of platinum, suitable palladium or rhodium compounds can be used as catalysts. If olefin complexes of platinum metal are used as a catalyst, crosslinking takes place at room temperature. For addition crosslinking at higher temperatures, platinum complexes modified with olefinic nitriles and acetylenic alcohols can be used[70–72]. The principle addition crosslinking reaction by hydrosilylation is shown in Figure 2.20. This technology is used on a large scale for solvent-free thermal curing release coatings and is described in the patent literature[73, 74].

Such products are commercially available from, for example, Bluestar Silicones, Dow Corning, Momentive and Shin-Etsu.

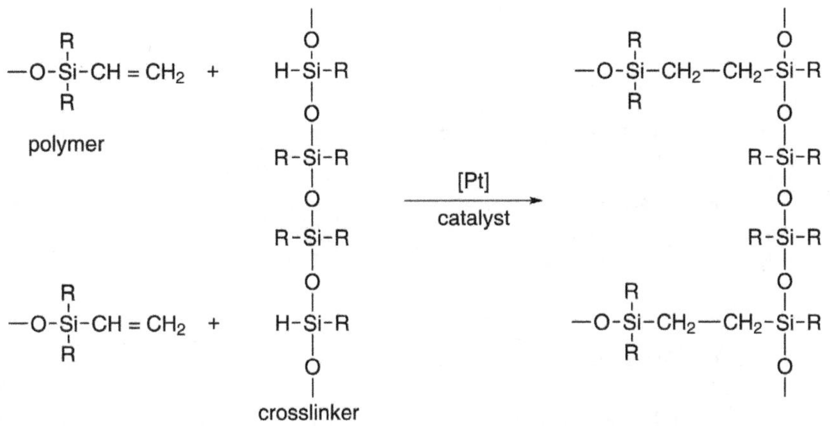

Figure 2.20: Addition crosslinking

For peroxide crosslinking, free radicals must first be formed. This can be induced either thermally or by radiation. Various organic peroxides are suitable as radical formers and act as initiators of this type of crosslinking (Figure 2.21). The incorporation of vinyl groups in the polymer (0.5 to 1.0 mol%) achieves more precise crosslinking and improved vulcanization.

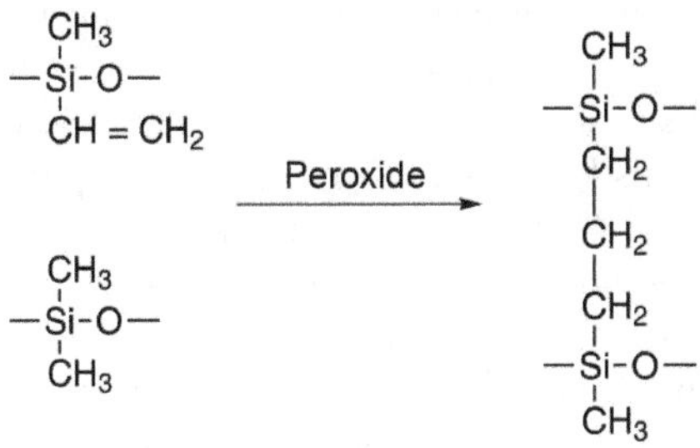

Figure 2.21: Crosslinking with peroxides

For condensation crosslinking, catalysts, especially tin soaps such as dibutyltin dilaurate are used to catalyze the reaction between dihydroxy polysiloxanes and silica esters (Figure 2.22). Water greatly accelerates the reaction rate which also depends on the crosslinker (functionality, concentration, chemical structure) and type of catalyst.

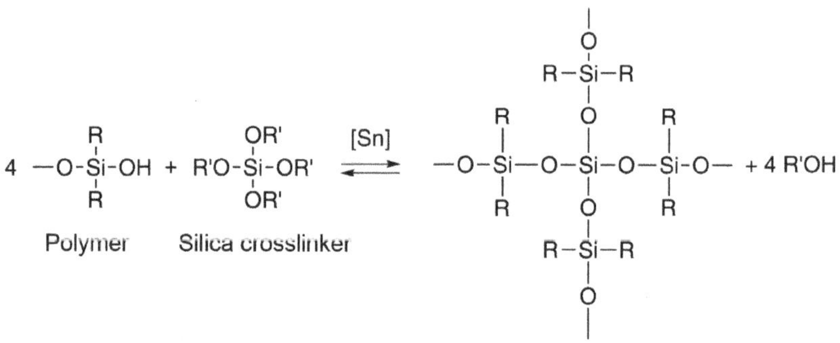

Figure 2.22: Condensation crosslinking

Such raw materials are available commercially from, for example, Dow Corning and Wacker.

2.6 Waterborne silicone resins

Waterborne silicone resins can be prepared both as primary and as secondary emulsions. Primary emulsions of silicone resins can be obtained via emulsion polymerization of sparingly soluble monomers such as octamethylcyclotetrasiloxane in combination with tri- or tetrafunctional alkoxy silanes: Monomers are added to an aqueous surfactant solution in combination with a strong acid or base as a catalyst. Particle growth occurs via diffusion of monomers through the aqueous phase and subsequent reaction within the surfactant micelles, resulting in stable siloxane polymer dispersions[75–77]. If large amounts of surfactant are used, the resulting particles are so small that the dispersion is translucent[78]. Alternatively, monomers are first emulsified using an emulsifier and then polymerized, a process known as mini-emulsion polymerization[79, 80]: polyreactions in mini-emulsions. The advantage of mini-emulsion polymerization over emulsion polymerization is that water-insoluble prepolymers such as α,ω-dihydroxy functionalized dimethylsiloxane and polymeric silicone resins can be used. The alcohols formed by the condensation of silanes or silicone resins are sometimes removed from the system by distillation to prevent the destabilization of the mini-emulsion through desorption of the emulsifier. Commercially available products typically have a high D-unit content (see Chapter 1.1) and are thus elastomeric in nature.

Of greater significance for coating applications are secondary emulsions in which solventborne silicone resins are converted into aqueous emulsions using any external emulsifier. A certain amount of solvent is often necessary as emulsification of highly viscous resins is difficult. Solvent removal by distillation of the final emulsion is impossible when high boiling solvents are used, and often not preferred as a residual amount of solvent is beneficial for proper coating formation. High solids silicone resins offer a compromise between solvent content, film formation and viscosity, resulting in a typical solvent content between 3 and 10 % of the emulsion. The choice of solvent also influences the emulsification process: solvents that are partially water miscible, such as methoxypropyl acetate will partition over both the silicone resin and aqueous phases and reduce the adsorption tendency of the emulsifier. Therefore it is not possible to prepare stable emulsions of resins from such solvents.

Secondary emulsions are available for methyl-and methyl/phenyl-silicone resins and combinations thereof. The stability of the emulsions is good. Though the silicone resin still contains reactive alkoxy groups, the presence of water alone does not lead to premature condensation if no catalyst is present. Waterborne silicone resins have established themselves primarily in building conservation for hydrophobization of mineral substrates, plasters and paints.

Microemulsions[81] are an interesting form of waterborne silicone resin. The particle size of these products is smaller than the wavelength of visible light so that they appear transparent.

The preparation of microemulsions is based on incorporation of amino-functional silanes in low molecular weight silicone resins. These resins act as a surfactant when the amino groups are protonated with acetic acid. They are blended with co-surfactants and other alkoxy-functional silicone resins to form water- and solvent-free concentrates. When adding water, a microemulsion forms spontaneously. This can be applied as a water-repellent primer. After evaporation of water and acetic acid, a water-insoluble silicone resin remains which reacts both with itself and with the mineral substrate. As hydrolysis of the alkoxy groups and condensation starts within a few hours, the lifetime of the diluted, aqueous microemulsion is limited[82]. In addition, penetration into the substrate diminishes with increasing molecular weight. Therefore microemulsions should preferably be used immediately after preparation. The solid content of such microemulsions is about 10 %. Significantly higher concentrations cannot be achieved as viscosity increases strongly above 20 %.

Microemulsions can also be prepared using trifunctional alkoxy silanes in a large excess of water and alcohol in the presence of emulsifiers and catalysts. Hardness can be improved by the use of silica sols which serve as highly crosslinked condensation nuclei for the silanes. Linear or weakly branched siloxanes with pendant amino or telichelic quaternary ammonium groups ("silicone quats") also form stable, transparent microemulsions in the presence of emulsifiers. Due to their linear character they are mainly used as glide additives in wood, textile and industrial coatings[83].

3 Examples of applications of silicone resins

Silicone resins feature in various fields:

A) Silicone resins for heat resistant coatings and corrosion protection above 300 °C
B) Silicone-modified aromatic polyester resins for heat resistant coatings up to 250 °C for decorative coatings
C) Application at room temperatures
- silicone-modified aliphatic epoxy resin for versatile coating applications
- acrylic- and epoxy-functional silicone resins for UV-cured release liners
- silicone acrylates as additives for UV-cured coatings.

D) Silicone resins in building conservation
- external water repellency
- internal water repellency.

3.1 Silicone resins for heat resistant coatings and corrosion protection above 300 °C

As already indicated in Chapters 2.1.1 and 2.1.2, silicone resins can be divided into methyl-silicone resins and phenyl-silicone resins.

The similarity of methyl-silicone resins to silica is the origin of the partially inorganic character of this group of resins. This explains properties such as the relatively high hardness, low thermoplasticity, poor pigmentability, and outstanding colour stability over a wide range of temperatures (up to 650 °C), particular affinity to inorganic, mineral products and incompatibility with organic resins. The methyl-silicone resins exhibit excellent water repellency even in a solvent-free, partially crosslinked state at room temperature. A further advantage is that the organic portion of methyl-silicone resin coatings decomposes on heating almost without producing any smoke.

W. Heilen, S. Herrwerth: Silicone Resins and their Combinations

ISBN: 978-3-86630-828-2

In contrast to methyl-silicone resins, methyl/phenyl-silicone resins show much better compatibility with organic resins and better pigment affinity. Furthermore they make less demands on substrate pretreatment.

In principle, methyl-silicone resins and methyl/phenyl-silicone resins can be divided into two groups: ambient curing systems and systems which cure at a higher temperature (baking systems). The significant structural difference between the two systems is shown in Figure 2.5 (see Chapter 2.1.1).

Suitable catalysts such as tetra n-butyl titanate (TnBT) or mixtures of this with tetramethylguanidine (TMG) are used to accelerate curing of ambient systems (see Table 3.1) in the presence of moisture. Atmospheric moisture is an important factor without which no curing occurs (see Chapter 2.1.1).

The ensuing high crosslink density in ambient curing systems leads to the resulting coatings being more resistant at temperatures up to 300 °C to solvents, oil, atmospheric moisture and salt spray than coatings produced with higher temperature curing systems (baking systems). High temperature curing silicone resin systems are typically baked for 30 min at approximately 250 °C.

The main field of application of these coatings is corrosion protection at high temperatures, e.g. for mufflers, exhaust pipes, industrial furnaces, ovens, barbecues, generators and turbines (see Figure 3.1).

The property profile of coatings based on methyl-silicone resins or methyl/phenyl-silicone resins is largely determined by the formulation[84]. For heat resistance up to 350 °C formulations with heat resistant inorganic coloured pigments (such as cobalt green, cobalt blue, nickel titanate yellow, red iron oxide, spinel black and titanium oxide) can be used. Even at continuous exposure to temperatures of 350 °C, the films are completely functional but gradually lose their gloss.

Metallic pigments such as aluminium flakes, iron mica and zinc dust enhance the heat resistance up to 650 °C. Formulation with flaky fillers (e.g. mica or talc at

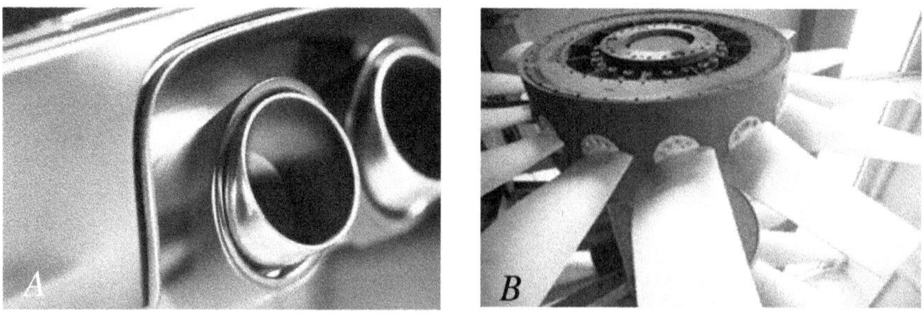

Figure 3.1: Typical fields of application for silicone resins

Table 3.1: Formulation based on an ambient temperature curing methyl/phenyl-silicone resin for heat resistance coatings up to 550 °C

1. Mill base			
Pos.	**Components**	**p.b.w.**	**Trademark holder**
1	"Silikophen" AC 900	40.50	Evonik Industries AG or one of its subsidiary companies
2	"Tego" Airex 900	0.50	Evonik Industries AG or one of its subsidiary companies
3	"Tego" Dispers 670	1.50	Evonik Industries AG or one of its subsidiary companies
4	"Bentone" SD-1	1.50	Elementis Specialties, Inc.
5	Mica TM	16.00	Aspanger Bergbau und Mineralwerke GmbH & Co KG
6	"Heucodur" Black 9-100	8.00	Dr. Hans Heubach GmbH & Co.KG
7	"Heucophos" ZPO	5.00	Dr. Hans Heubach GmbH & Co.KG
8	"Aerosil" 200	1.50	Evonik Industries AG or one of its subsidiary companies
9	Xylene	13.00	
Total mill base		87.50	
2. Let down			
Pos.	**Components**	**p.b.w.**	**Trademark holder**
10	Butyl glycol acetate	2.10	
11	Xylene	10.40	
Total		100.00	
Manufacture			
Position 1	Weigh in.		
Position 2–9	Add while stirring. Disperse in the bead mill until the required particle size is reached (< 20 µm)		
Position 10–11	Add components while stirring and adjust to application viscosity.		
Mixing ratio			
Hardener mixture	"Tego" Kat 1[1]: "Tego" Kat 2[1]= 1 : 1		
Coating system	Coating : Hardener mixture= 100 : 0.6		
[1] Evonik Industries AG or one of its subsidiary companies			
Coating properties			
Attribute		Value	Unit
Theoretical solid content		approx. 70	%
Cup efflux time (Cup: DIN 6)		23 ± 5	s
Density		1.28	g/cm³
VOC-content		approx. 380	g/l

amounts greater than 15 %) is important in imparting some flexibility to the coating in alternating temperature stress tests.

At temperatures above 350 °C organic groups of the resin are almost completely volatilized. Aided by sintering, very hard inorganic composite systems are gradually formed. The organic groups, particularly the phenyl substituents, leave small voids when burnt off. The composite formed is still flexible so that coatings at a temperature of 600 °C can be repeatedly quenched with cold water without cracking.

Coatings resistant to temperatures above 700 °C can be obtained by using ceramic powders (frits). However, the entire coated surface must be heated to temperatures above the melting point of the frits. Enamel-like coatings are obtained, which are bonded firmly to the siloxane skeleton.

The Pigment Volume Concentration (PVC) also plays a role in the flexibility of the formulation. It should be between 20 and 30 %.

To achieve long-term corrosion protection a zinc-dust primer must be applied preferable based on silicone-modified aromatic epoxy resins (see Chapter 2.2.3) with a temperature resistance up to 450 °C.

The dry film thickness of such coatings should not exceed 30 to 40 µm otherwise brittleness or delamination may occur after long term exposure to heat.

The pretreatment of the substrate significantly affects adhesion and protective properties.

In general the substrate should be free from oil, grease, mill-scale and welding residues. For the methyl-silicone resins the steel substrates must be shot-blasted. For the methyl/phenyl-silicone resins the substrate pretreatments are less rigorous than for the methyl-silicone resins. The substrates should be shot-blasted. Methyl/phenyl-silicone resins can also be used for untreated steel substrates.

Phosphated substrates cannot be used in combination with a high temperature resistant coating based on silicone resins, because of delamination caused by water of crystallization in the phosphated substrates.

Requirements to reduce solvent emissions (VOC) can be achieved by using either high solids systems provided by, for example, ambient-curing resins or waterborne silicone resin emulsions. Coatings based on these waterborne silicone resin emulsions have properties corresponding to those of solventborne silicone resins (see Table 3.2).

It should be pointed out that metals (e.g. tin) from packaging drums which come into contact with silicone resins can cause the formation of gel particles. It is therefore recommended that formulated coatings should not be stored or filled into such containers.

Table 3.2: Formulation based on a waterborne silicone resin emulsion for heat resistant coatings up to 500 °C

1. Mill base			
Pos.	**Components**	**p.b.w.**	**Trademark holder**
1	Water, demin.	20.10	
2	"Tego" Foamex K3	0.20	Evonik Industries AG or one of its subsidiary companies
3	"Tego" Dispers 750 W	1.50	Evonik Industries AG or one of its subsidiary companies
4	NH_4OH (24.5 %)	0.20	
5	"Heucodur" Black 9-100	8.00	Dr. Hans Heubach GmbH & Co.KG
6	"Heucosil" CTF	5.00	Dr. Hans Heubach GmbH & Co.KG
7	Mica TM	14.00	Aspanger Bergbau und Mineralwerke GmbH & Co KG
Total mill base		49.00	
2. Let down			
Pos.	**Components**	**p.b.w.**	**Trademark holder**
8	"Silikophen" P 40/W	45.00	Evonik Industries AG or one of its subsidiary companies
9	Water, demin.	6.00	
Total		100.00	
Manufacture			
Position 1	Weigh in.		
Position 2–7	Add while stirring. Disperse in the bead mill until the required particle size is reached (< 20 µm).		
Position 8	Add while stirring.		
Position 9	Add while stirring and reduce to application viscosity.		
Coating properties			
Attribute		**Value**	**Unit**
Theoretical solid content		approx. 50	%
Cup efflux time (Cup: DIN 6)		40	s
pH-Value (10 % sol.)		approx. 8.7	
Density		1.25	g/cm³
VOC-content		approx. 290	g/l

3.2 Silicone-modified aromatic polyester resins for heat resistant coatings up to 250 °C for decorative coatings

As described in Chapter 2.2.1 the silicone-modified aromatic polyester resins are produced by the reaction of OH-functional aromatic polyesters, based on aromatic dicarboxylic acids (e.g. isophthalic and terephthalic acid), with alkoxy-functional silicone intermediates. These resins can withstand long-term exposure to temperatures of 250 °C. Polar solvents, including small quantities of alcohols, are used to prevent undesirable side reactions and to maintain a stable viscosity during resin manufacture. By varying the polyester building blocks and the silicone resin intermediates, it is possible to manufacture silicone polyester resins with differing properties.

The heat resistance properties of the silicone-modified aromatic polyester resins can be tailored by the silicone content and the choice of aromatic polyester resin. The amount of silicone in this type of resin can vary between 10 and 80 % w/w. Heat resistance improves with increasing silicone content. The silicone component also provides gloss retention and better weathering resistance. Generally, the presence of the polyester is intended to provide more favourable curing conditions, greater flexibility, better compatibility with other binders and enables good pigment wetting. For example, it is possible to manufacture white coatings with degrees of gloss of up to more than 90 gloss units measured at a geometry of 20°.

Formulation of light colours, in particular, is feasible because coatings based on silicone-modified aromatic polyester resins have a low susceptibility to thermally-induced yellowing.

A further characteristic of coatings based on high quality silicone-modified aromatic polyester resins is their low thermoplasticity. Since the silicone component

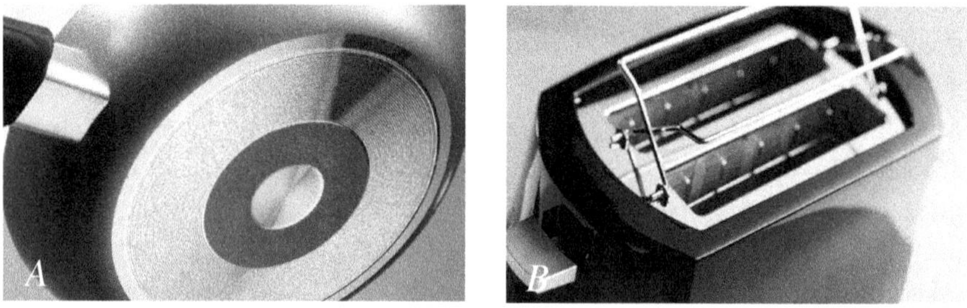

Figure 3.2: Typical fields of application for decorative coatings based on silicone polyester resin

Table 3.3: Formulation based on silicone-modified aromatic silicone polyester with a silicone content of approx. 80 %

1. Mill base			
Pos.	**Components**	**p.b.w.**	**Trademark holder**
1	"Silikoftal" HTT	34.70	Evonik Industries AG or one of its subsidiary companies
2	"Blanc fixe" micro	3.10	Sachtleben Chemie GmbH
3	"Kronos" 2360	31.10	Kronos Titan GmbH
4	"Aerosil" R 972	0.50	Evonik Industries AG or one of its subsidiary companies
Total mill base		69.40	
2. Let down			
Pos.	**Components**	**p.b.w.**	**Trademark holder**
5	"Silikoftal" HTT	22.30	Evonik Industries AG or one of its subsidiary companies
6	Methoxypropyl acetate	5.20	
7	Butyl glycol acetate	1.00	
8	Butyl acetate	2.10	
Total		100.00	
Manufacture			
Position 1	Weigh in.		
Position 1	Weigh in.		
Position 2–4	Add while stirring. Disperse in the bead mill until the required particle size is reached (< 20 µm)		
Position 5–8	Add components while stirring.		
Attribute		**Value**	**Unit**
Theoretical solid content		approx. 77	%
Cup efflux time (Cup: DIN 6)		30–40	s
Density		1.54	g/cm³
VOC-content		approx. 350	g/l

and the polyester component are both strongly crosslinked, hardness remains high even at temperatures above 150 °C. This is important in applications where the hot coatings are mechanically stressed and the surface must not suffer damage.

The properties of these resins are exploited in decorative, heat resistant coatings for domestic electrical appliances such as toasters, sunlamps, fan heaters, electric iron soles, cookers and external coatings of deep fryers, bakeware, pots and pans (see Figure 3.2).

An important property of coatings used in household applications is detergent resistance. This characterizes the ability to remain unaffected by long-term exposure to surfactant-based and strongly alkaline cleaning agents in dishwashers. The detergent resistance of a coating is generally determined by its formulation but also by the binder used. It is thus an important quality criterion for silicone polyesters.

Points to note when processing silicone-polyester resins

The silicone-modified aromatic polyester resins are intrinsically polar: A very high level of OH groups is present which, together with the ester bonds, can generate strong interactions. These may even falsely indicate high molecular weights in gel chromatograms. As previously described, polar solvents and small amounts of alcohols are used in the manufacture of the resins to prevent association and stabilize the viscosity.

Polar solvents, such as ketones or esters, should also preferably be used in the processing of coatings since they prevent turbidity e.g. caused by low storage temperatures or extended storage. This cloudiness, which does not otherwise impair the quality of the coating, stems from small amounts of polyester which have not reacted during the modification reaction. Aliphatic solvents should not be used. Volatile alcohols should only be used in blends with other solvents since otherwise the stoichiometry of the stoving reaction is altered. The use of high boiling alcohols can cause significant deterioration in the thermal stability of the coating film in terms of yellowing resistance.

Modern silicone polyester systems exhibit a significantly higher acceptance of aromatic solvents. The advantages lie primarily in the degree of freedom in formulating to optimize conductivity for electrostatic application.

OH group content

As with pure silicone resins, it is not possible to give an exact figure for OH content as there is no certain and simple method of determining the polymer-bound OH groups independently of the OH groups from alcohols which have been formed or subsequently added. During manufacture, attention is paid to viscosity and transparency rather than to a constant hydroxyl number. These criteria depend however on the polymer starting products, variations in conversion and thus somewhat differing OH content.

Stoving conditions and catalysts

Generally, the curing reactions of silicone polyester resins can be accelerated by catalysts such as acids, tin and titanium compounds. There are, however, upper limits on the amount of catalyst used to lower stoving temperatures significantly since problems such as impaired storage stability or even gelling, embrittlement or increased susceptibility to yellowing can occur.

For complete curing without catalysts, the following conditions are recommended:

60 minutes/200 °C

30 minutes/250 °C

10 minutes/270 °C

Crosslinkers, such as melamine resins, can also be used to alter the crosslinking density allowing tailoring of the final coating properties. Care must be taken to ensure that the crosslinkers do not impair the heat resistance.

3.3 Applications at normal temperatures

3.3.1 Silicone-modified aliphatic epoxy resin for versatile coating applications

Silicone-modified aliphatic epoxy resins can be used as solvent-free, ultra high solids binders for resource-conserving, multifunctional industrial top coat formulations in many different applications. This technology allows formulation of coating systems with significantly less than 250 g/l (sometimes even less than 100 g/l) VOC as well as permitting isocyanate-free crosslinking.

Noteworthy properties of coatings based on silicone-modified aliphatic epoxy resin technology are: outstanding anti-corrosion properties, high weather resistance, high chemical resistance, good pigmentability, low flammability, good stain resistance, high hardness, excellent abrasion resistance, easy-to-clean effect and good adhesion[21].

The multiplicity of favourable properties of silicone-modified aliphatic epoxy resins originates from the combination of the three chemical building blocks, aliphatic epoxy resin, silicone resin and amino-trialkoxy silane, together with the high crosslinking density originating from the dual-cure mechanism (see Figure 2.13). As described in Chapter 2.2.3 the amino groups of the amino-trialkoxy silane react with the epoxy groups of the silicone-modified epoxy resin. The alkoxy groups of the amino-trialkoxy silane then react by hydrolyzing/condensing in the presence of atmospheric moisture with the free alkoxy groups of the polysiloxane. Due to the significant molar excess of alkoxy groups, the hydrolysis/condensation reaction can be catalyzed by Lewis acids, acids, or amines and drying is consequently accelerated. Acceleration caused by the use of the Lewis acid, dioctyltindineodecanoate, as a catalyst is shown in Figure 3.3. Consequently the curing rate of modified aliphatic epoxy resins can be accelerated to approximately that of two-pack polyurethane coatings by the use of suitable catalysts.

Table 3.4: High gloss two-pack top coat formulation with a solids content of approx. 95 % and a VOC content of approx. 70 g/l based on silicone-modified aliphatic epoxy resin

1. Mill base			
1	"Silikopon" EF	30.00	Evonik Industries AG or one of its subsidiary companies
2	"Tego" Airex 900	1.00	Evonik Industries AG or one of its subsidiary companies
3	"Tego" Dispers 670	1.00	Evonik Industries AG or one of its subsidiary companies
4	"Thixatrol" ST	1.00	Elementis Specialties, Inc.
5	"Kronos" 2360	30.50	Kronos Titan GmbH
6	"Blanc fixe" micro	7.00	Sachtleben Chemie GmbH
7	Xylene	1.00	
Total mill base		71.50	
2. Let down			
Pos.	**Components**	**p.b.w.**	**Trademark holder**
8	"Silikopon" EF	25.00	Evonik Industries AG or one of its subsidiary companies
9	"Tinuvin" 292	1.00	BASF SE
10	Butyl acetate	2.50	
Total		100.00	
Manufacture			
Position 1	Weigh in.		
Position 2–7	Add while stirring. Disperse in the bead mill until the required particle size is reached (< 20 µm)		
Position 8–10	Add components while stirring.		
Mixing ratio			
Hardener mixture	"Dynasylan" AMEO[1] : "Jeffamine" D-230[2]=6.9:1		
Coating system	Coating : Hardener mixture=100:11.6		
To accelerate the curing 3% of "Tib Kat" 318[3] (calculated on solid resin) can be used.			

[1] Evonik Industries AG or one of its subsidiary companies
[2] Huntsman Corporation or an affiliate thereof Value Unit
[3] TIB Chemicals AG approx. 70 %

Coating properties		
Attribute	**Value**	**Unit**
Theoretical solid content	approx. 95	%
Cup efflux time (Cup: DIN 6)	35–45	s
Density	1.49	g/cm³
VOC-content	approx. 70	g/l

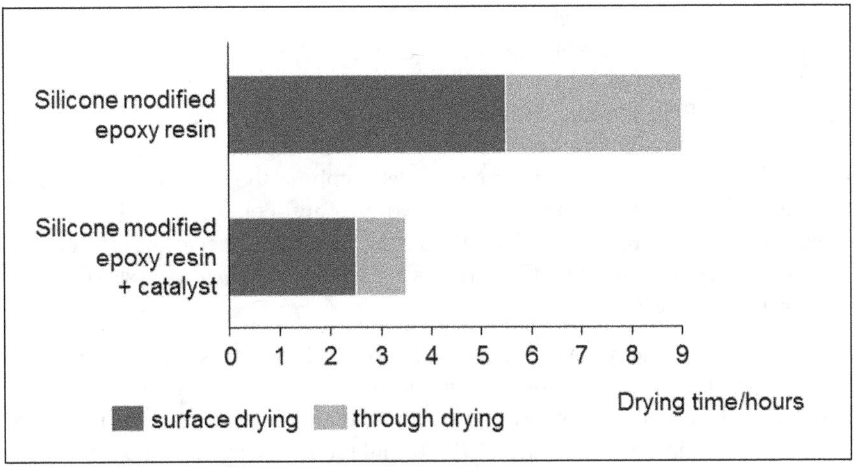

Figure 3.3: Influence of catalysis on curing speed using 2 % dioctyltindineodecanoate

Silicone-modified aliphatic epoxy resins have found wide application particularly in corrosion protection for which the coating must meet a whole range of requirements to be effective. It must protect the underlying steel by adhering well while simultaneously preventing harmful substances coming into contact with it. The coating must, itself, be weather resistant. This means not only resistant to rain, snow and ice but also, for example, to the high energy UV component of sunlight. Often, additional active protection of the steel, with for example a zinc primer, is necessary.

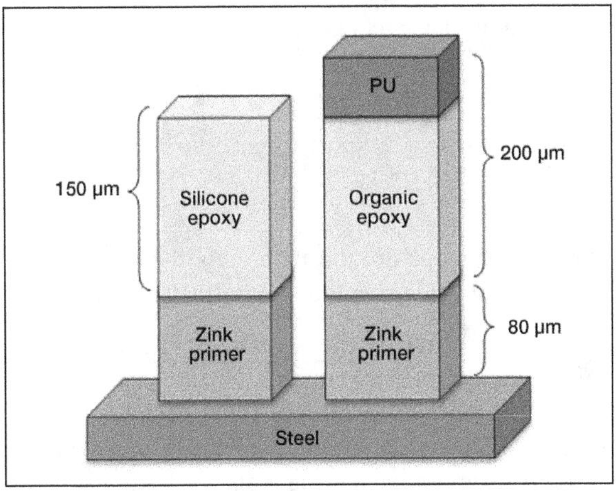

Figure 3.4: Comparison of a conventional three-coat structure and a silicone-modified aliphatic epoxy resin based two-coat structure for heavy duty anti-corrosion coatings

Conventional coating systems for achieving Category C5 heavy-duty corrosion protection (as used in maritime environments, particularly in off-shore applications) are generally based on a three-coat structure. This is composed of a two-pack epoxy zinc-dust primer, a two-pack epoxy intermediate coat and a two-pack polyurethane (PUR) top coat. Generally the individual coats are 80 µm, 150 µm and 50 µm thick, respectively, leading to a total thickness of around 280 µm[21]. As the epoxy intermediate coat is based on aromatic epoxy resins it exhibits weakness in its weathering resistance and must of necessity be protected by a two-pack PUR top-coat. Silicone-modified aliphatic epoxy resins with their tailored chemical combination of inorganic resin blocks (the silicone resins) with weathering-resistant aliphatic epoxy resins combine the properties of primer and top coat. Using these, new types of two-coat anti-corrosion coatings can be produced which are thinner but have the same protective effect as the thicker conventional three-coat systems. This allows formulation of coating systems which are sparing with resources.

Further, using the isocyanate-free silicone-modified epoxy resins, dry film thicknesses of 150 µm can be achieved without formation of reaction bubbles. This is not usually possible with two-pack polyurethane coatings in one operation. Bubble formation with two-pack PUR formulations occurs from possible side-reactions of the isocyanate with atmospheric moisture.

In the most recent investigations it has been found that using silicone-modified aliphatic epoxy resins a further reduction in the dry film thickness from 150 down to 100 µm is possible without impairing the anti-corrosion properties of the two-coat structure.

Furthermore, as shown in Table 3.5, silicone-modified epoxy resin technology allows manufacture of direct-to-metal coatings with excellent anti-corrosion properties.

Silicone-modified aliphatic epoxy resins are not only used for anti-corrosion coatings but also in many other applications, such as

- concrete coatings
- flooring applications
- easy-to-clean coatings
- foul release coatings
- wood coatings.

Advantages in using such coatings for wood are the flame retardant properties, high scratch resistance and the brilliant grain accentuation of the surface.

As coatings based on silicone-modified aliphatic epoxy resins have a high resistance to chemicals they are also suitable for floor and anti-graffiti applications. After removing the adhering graffiti with anti-graffiti cleaner based on strong solvents and surfactants, no residual soiling or changes in gloss or colour are apparent. This permits new ways of formulating soiling-resistant coatings.

Their robust and easy-to-clean properties permit the formulation of foul release coatings for marine applications.

Table 3.5: Two-pack direct-to-metal formulation with a solids content of approx. 85 % and a VOC content of approx.150 g/l based on silicone-modified aliphatic epoxy resin

1. Mill base			
Pos.	**Components**	**p.b.w.**	**Trademark holder**
1	"Silikopon" EF	30.00	Evonik Industries AG or one of its subsidiary companies
2	"Tego" Airex 900	1.00	Evonik Industries AG or one of its subsidiary companies
3	"Tego" Dispers 670	0.80	Evonik Industries AG or one of its subsidiary companies
4	"Thixatrol" ST	1.00	Elementis Specialties, Inc.
5	"Kronos" 2315	27.50	Kronos Titan GmbH
6	"Blanc Fixe" micro	3.00	Sachtleben Chemie GmbH
7	"Heucophos" ZAM-Plus	3.00	Dr. Hans Heubach GmbH & Co.KG
8	Xylene	1.00	
Total mill base		67,30	
2. Let down			
Pos.	**Components**	**p.b.w.**	**Trademark holder**
9	"Silikopon" EF	14.50	Evonik Industries AG or one of its subsidiary companies
10	"Degalan" LP65/11 (45 % solution in butyl acetate)	11.00	Evonik Industries AG or one of its subsidiary companies
11	"Tinuvin" 292	1.00	BASF SE
12	Butyl acetate	6.20	
Total		100.00	
Manufacture			
Position 1	Weigh in.		
Position 2–9	Add while stirring. Disperse in the bead mill until the required particle size is reached (< 20 µm)		
Position 10–11	Add components while stirring and adjust to application viscosity.		
Mixing ratio			
Hardener mixture	"Dynasylan" AMEO[1]: "Jeffamine" D-230[2] = 6.9 : 1		
Coating system	Coating : Hardener mixture= 100 : 9.4		
To accelerate the curing 3 % of "Tib Kat" 318[3] (calculated on solid resin) can be used.			
Coating properties			
Attribute		**Value**	**Unit**
Theoretical solid content		approx. 85	%
Cup efflux time (Cup: DIN 6)		35 45	s
Density		1.38	g/cm³
VOC-content		approx. 210	g/l

[1] Evonik Industries AG or one of its subsidiary companies; [2] Huntsman Corporation or an affiliate thereof

In concrete coating, chemical bonding occurs between silicone-modified aliphatic epoxy resins with amino-trialkoxy silanes and hydroxyl groups of the concrete substrate. This leads to a very strong coating on the concrete which cannot be removed mechanically.

Currently, the most common substrates for these silicone-modified aliphatic epoxy resins are metal and concrete. Coating of other substrates such as wood and glass are also of interest.

Coatings based on silicone-modified aliphatic epoxy resins can be used to protect a wide range of structures (see Figure 3.5):

- bridges
- exterior/interior of storage tanks
- drilling platforms
- oil pipelines
- ship decks
- underwater structures (foul release coatings)
- steel structures
- concrete floors and walls such as bridges and tunnels
- wood decks of sport boats.

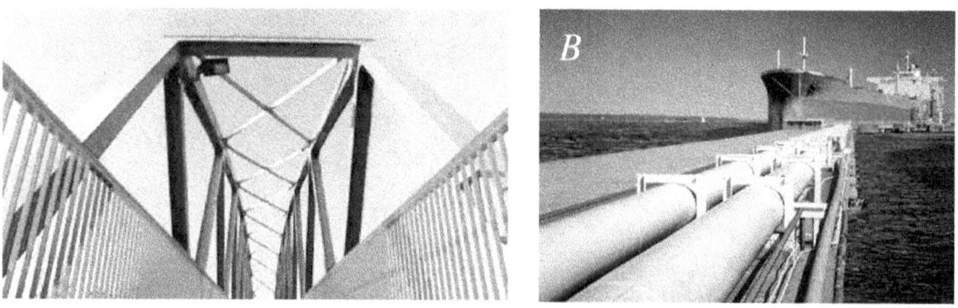

Figure 3.5: Typical field of applications for silicone-modified aliphatic epoxy resins

Formulation hints for silicone-modified aliphatic epoxy resins based coatings

- **Pigments and fillers**

The choice of a suitable titanium dioxide for white formulations can be difficult. Parameters such as storage stability, gloss retention, colour stability and hiding power must be assessed. The colour stability of organic pigments brightened with titanium dioxide must be evaluated after weathering. Moreover the adhesion and flexibility of the coating are particularly affected by the fillers used.

- **Solvents**

The choice of a suitable solvent for thinning silicone-epoxy hybrid coatings does not pose a problem. It is only necessary to take care that the water content is suitably low since water leads to unwanted polycondensation of the remaining methoxy groups from the silicone-epoxy hybrid resin. Small amounts of butanol or butyl acetate are ideal. Because of the low viscosity of the curing agents, further addition of solvent is unnecessary in most cases.

When using additives the formulator should always bear in mind the final properties of the film. In most cases, light stabilizers and a deaerator suffice. In individual cases, a flow control agent can ensure a smooth finish of extremely high quality.

3.3.2 Acrylic- and epoxy-functional silicone resins as UV-cured release coatings

The principle construction of a label laminate with the release coating on the substrate (backing paper) on one side and the label with the adhesive on the other side is shown in Figure 3.6. Label laminates are the main market for silicone release coatings today.

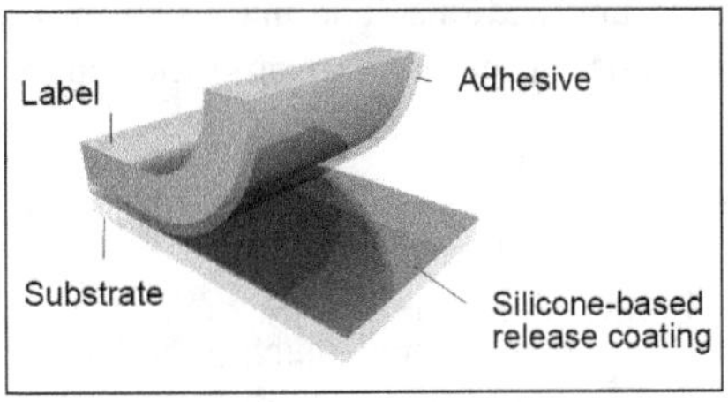

Figure 3.6: Principle construction of a label laminate

Technological development of silicone release coatings for the PSA (pressure sensitive adhesive) market began in the nineteen fifties with solventborne RTV (room temperature vulcanizing) resins. The curing process of this first generation is based on tin-carboxylate catalyzed condensation crosslinking (see Chapter 2.5 and Figure 2.22) of oligohydroxy-polysiloxanes and silica esters. Alkoxy- or hydrogen-functional polysiloxanes can also be used in this curing method. The by-products of the condensation reaction can be alcohols, carboxylic acids or hydrogen.

Evolving environmental concerns soon resulted in demand for solvent-reduced products.

Solvent-less RTV resins based on SiH-functional and Si-vinyl-functional polysiloxanes were developed in the late nineteen sixties. The curing process is based on heavy metal, e.g. platinum or rhodium, catalyzed addition crosslinking of the SiH-functional polysiloxanes with vinyl-functional polysiloxanes (see Chapter 2.5 and Figure 2.20). Because of the need for fast curing processes, temperatures above 100 °C are used to accelerate crosslinking. These drying conditions and fast production processes require very large curing ovens. Up to now this technology has been used on a large scale but the high temperature required for fast addition-crosslinking has several disadvantages:

- A change in the physical properties of the backing papers of the release liners and a significant reduction of the natural moisture content of the paper. This loss of moisture has to be compensated by cost-intensive re-humidification processes. The paper often shows poor lay-flat behaviour, also known as curling, caused by exposure to heat.
- A limitation on the use of plastic films, because of their heat sensitivity.

Solvent-less radiation-curable acrylic- and epoxy-functional silicone resins obviate the disadvantages of thermal curing silicones described above. The development of these low-emission and cold-curing systems started in the nineteen eighties with fast growing popularity in the last 25 years. The market share is expected to grow further.

Today the acrylic- and epoxy-functional silicone resins cover the whole range of release levels (premium/easy/controlled/tight release applications). There are complete systems based on various acrylic- and epoxy-functional silicone resin components which can be mixed with each other to adapt the release characteristic to satisfy the requirements of specific applications. The manufacture of commercially used acrylic- and epoxy-functional silicone resins and the patent literature are described in Chapter 2.4.

UV curing of the acrylic- and epoxy-functional silicone resins requires the addition of photoinitiators.

The acrylic-functional silicone resins cure via a free radical mechanism using photoinitiators e.g. 2-hydroxy ketones. When irradiated with UV light, the pho-

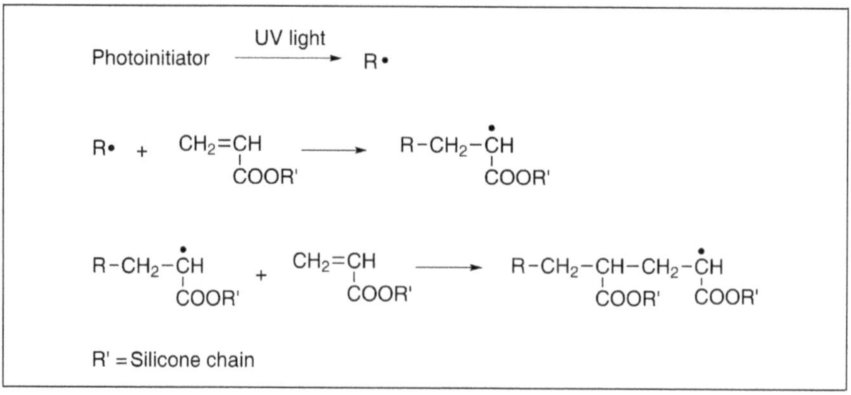

Figure 3.7: Free radical polymerization mechanism by UV-light

toinitiators are promoted to an excited state. They form radicals which initiate a chain reaction by attaching to CC double bonds to form new reactive radicals which, in turn, attach themselves to further CC double bonds, leading finally to a highly crosslinked three dimensional network (Figure 3.7).

In principle acrylic-functional silicone resins can also be cured by electron-beam radiation without needing a photoinitiator. However electron-beam curing is of minor commercial interest today due to high investment costs.

The epoxy-functional silicone resins cure in the presence of a cationic photocatalyst, e.g. iodonium-salts (Figure 3.8).

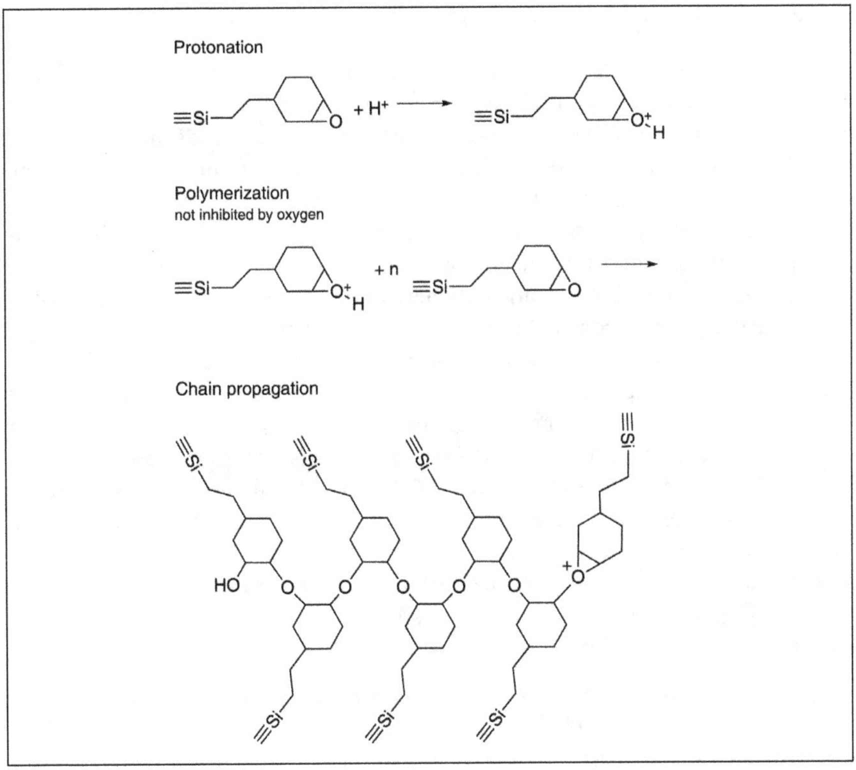

Figure 3.8: Polymerization of epoxy-functional silicone resins by UV-curing with a cationic photocatalyst

The free radical polymerization of the acrylic groups is much faster than cationic polymerization (Figure 3.9) of the epoxy groups. Therefore, full cure is reached more quickly with acrylic-functional silicone resins. The UV curable acrylic-functional silicone resins are very robust and cure is unaffected by, for example, impurities in the substrates. This allows many substrates to be siliconized, especially paper substrates. There is no post-curing and no photoinitiator poisoning as with epoxy-functional silicone resins. An additional advantage is that formulations with acrylic-functional silicone resins have a very long pot-life. The main disadvantage of radical curable systems is the need to inert the coating unit with nitrogen because the presence of oxygen will terminate polymerization. For the acrylic-functional silicone resins this is somewhat technically demanding. In general it is necessary to reduce the oxygen content to less than 50 ppm in thc UV unit to obtain an acceptable cure of the coating. With acrylic-functional silicone resins, use of special phosphite additives allows oxygen content to be increased up to 800 ppm in the UV unit without affecting curing quality. State of the art, inerted UV units are offered by a number of manufacturers.

Cationic curing takes place in a normal atmosphere and therefore, nitrogen inerting is not required. The cationic systems can be cured with standard UV equipment as used in the printing industry. This is the cationic system's main advantage over free radical curing silicone resins. Cationic UV siliconizing is therefore often the first choice when standard release values are required on suitable substrates (i.e. substrates which contain no ingredients that may poison the catalyst). Poisoning by humidity and substrates as well as post-curing are intrinsic to cationic curing. The cure speed of cationic polymerization is generally slower than that of free radical curing silicones. After exiting the UV source area, cationic polymerization continues in a post-curing process without further UV irradiation. The post-curing process of the cationic curing makes the handling of the received release liners much more complicated, because the release values change over the next 24 h after the UV-curing process. Also silicone transfer from the release liner to the adhesion layer (see Figure 3.6) due to insufficient cure can result in low release values and reduced adhesive tack of a label laminate construction. Figure 3.9 illustrates curing speed time profiles for radical and cationic curable systems. The silicones used in Figure 3.9 have the same degree of functionality[85]. The curing speed of the epoxy-functional silicone resins can be accelerated by the use of sensitizers, e.g. isopropylthioxanthone (ITX).

In-line corona treatment of the substrate just before siliconizing is always recommended when using free radical-curing silicones. It can also help improve anchoring of cationic-curing silicones but, in general, epoxy-functional silicone resins show good adhesion on different substrates. Corona treatment forms hydroxyl, carboxyl, and free radical groups, which are important for anchoring acrylic-functional silicone resins.

Solvent-free, cold cure of epoxy- and acrylic-functional silicone resins allows the use of heat-sensitive substrates, e.g. PE and PP films, and minimization of substrate film thickness. Paper substrates do not need to be re-moisturized and

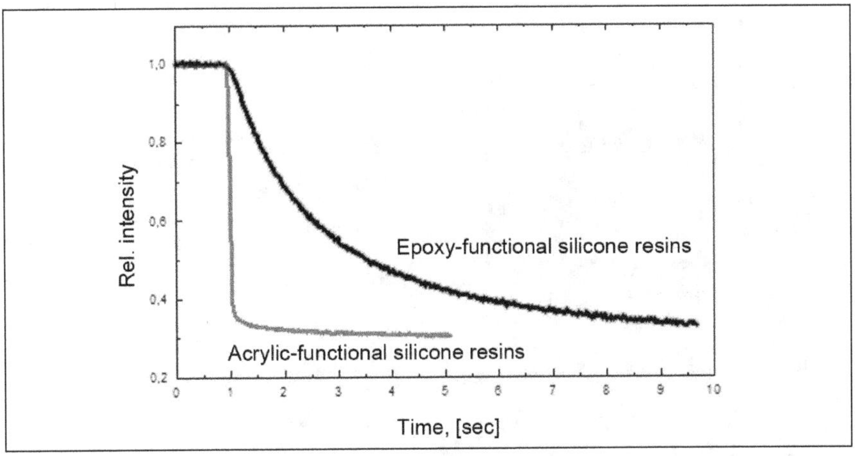

Figure 3.9: Curing speed profiles for radical and cationic curable systems. Real time IR measured at 1407 cm^{-1} (double bond) and 1068 cm^{-1} (ether).

show absolute lay-flat performance. The very fast cure makes it possible to use UV-curing silicone resins in inline processes together with adhesive coating. It also permits very high line speeds. The UV curing equipment is very compact. Compared with traditional thermal ovens, it saves space and energy. Typical fields of application of release liners are self-adhesive labels, graphic arts, tapes and hygienic products.

3.4 Silicones and silicone resins in building conservation

3.4.1 External water repellency

Façades can be rendered water repellent in various ways. External waterproofing of building materials (impregnation) can reduce water pick-up[86]. Impregnation is certainly the most effective way of protecting building materials which will not be further coated but it is not suitable for those which have low absorbing power.

High wall moisture can only be effectively combated if the different mechanisms of water absorption are considered (Figure 3.10).

A very common cause lies in poor building technique such as damaged joints, fissures, failing sealing and other errors in construction. These include wrongly designed window frames and wall projections and moldings which do not provide for free drainage of rain water. A frequent cause of failures is defective sealing of roofs and guttering. Such building defects lead to high rain-water pick-up. Subse-

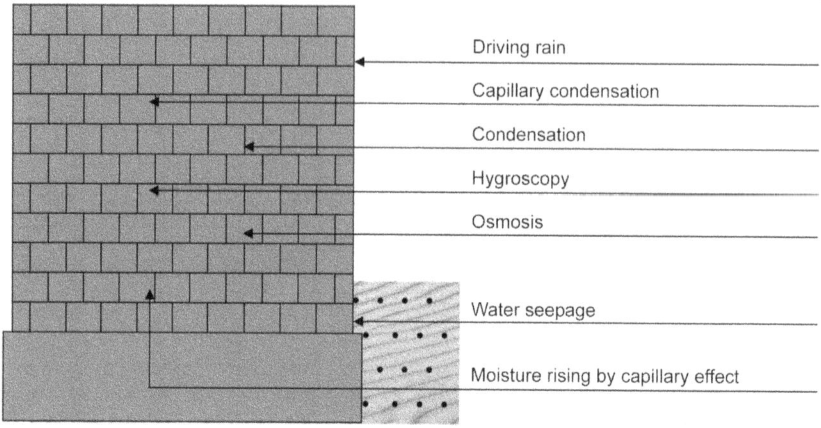

Figure 3.10: Mechanisms of water absorption

quent drying-out is hindered by hygroscopic salts. Moisture moves under osmotic pressure and makes the actual source of the damage difficult to determine.

One of the most important mechanisms for water pick-up is condensation of atmospheric moisture when the temperature of the cold surface falls below the dew point. However, even above the dew point condensation occurs in the capillaries of the building material. In the lower parts of walls there is often ground water pressure adding to moisture caused by capillaries.

The most conspicuous form of damage caused by moisture is efflorescence of salts. Water-soluble salts from bricks and joints reach the surface, the water evaporates and the salts crystallize out.

Lime- and cement-bonded building materials are leached by water and CO_2 forming soluble bicarbonates (1). Even without air pollution, easily soluble alkali sulfates (which are now used in much greater quantity in bricks and tiles than previously) form calcium sulfate (2) and absorption of water of crystallization produces rupture pressure (3). Release of CO_2 and water produces easily soluble sodium carbonate which effloresces (4). SO_2, in a multi-step reaction catalyzed by oxides of nitrogen, leads to the formation of sodium sulfate (5) which causes both pressure from crystallization (6) and further changes in lime-based cements. All reactions are shown in Figure 3.11. Sodium sulfate is one of the salts which are most damaging to buildings as the reversible uptake of water of crystallization leads to continuous fluctuation in crystallization in the warmer seasons. Above

$$1)\ CaCO_3 + CO_2 + H_2O \longrightarrow Ca(HCO_3)_2$$

$$2)\ Ca(HCO_3)_2 + Na_2SO_4 \longrightarrow 2\ NaHCO_3 + CaSO_4$$

$$3)\ CaSO_4 + 2\ H_2O \rightleftharpoons CaSO_4 \cdot aq.$$

$$4)\ 2\ NaHCO_3 \longrightarrow Na_2CO_3 + H_2O + CO_2$$

$$5)\ Na_2CO_3 + SO_2 + \tfrac{1}{2}\ O_2 \longrightarrow Na_2SO_4 + CO_2$$

$$6)\ Na_2SO_4 + 10\ H_2O \underset{> 32.7\ °C}{\overset{< 32.7\ °C}{\rightleftharpoons}} Na_2SO_4 \cdot aq.$$

Figure 3.11: Chemical reactions in lime- and cement-bonded materials

32 °C, sodium sulfate completely loses its water of crystallization whereas, under 32 °C, 10 mol of water of crystallization are taken up. This change in crystallization is far more dangerous than the effect of freezing/thawing in winter.

It should be noted at this point that the mechanisms of deterioration are not as simple as appears from the given reaction sequences. It would certainly be incorrect to assume that there would be no damage in the absence of wall dampness. The "dry" stone surface already contains many water molecules and can absorb considerable amounts of SO_2 from the air producing sulfurous acid which is partly neutralized to sulfites on the surface and partly oxidized to sulfuric acid and its salts. This water has a key role in the deterioration of mineral building materials.

Effective waterproofing must protect the coating against infiltration of salts and moisture and also improve its adhesion. In the event of damage to the top coat, the primer must protect the building material against further damage.

With low surface tension such as is obtained by impregnating surfaces with organo-siloxanes, water can be prevented from penetrating into pores and capillaries. At high angles of contact, that is in the region of capillary depression, water cannot rise in the capillaries – on the contrary it is expelled from them by negative capillary forces.

A valuable advantage of using silicones as external water repellents is that they lower dirt deposition along hairline cracks. Rain water, laden with dirt, running off from the façade is absorbed at the hairline cracks and the larger dirt particles stick in them. After some time, more and more dirt collects so that the hairline cracks can easily be seen. Waterproofing building material destroys its absorbency and reduces the visibility of the cracks.

Organic polymers are not suitable as impregnating agents as most of them have inadequate penetration to ensure long-term effectiveness. With a small depth of penetration, subsequent fine fissure formation causes rapid loss of effectiveness. If adequate penetration is achieved, these fissures lie within the water repellent zone and do not markedly alter water pick-up. Water vapour diffusion is strongly reduced by organic polymers and the breathability of the wall is strongly impaired. Additionally, the water repellency achieved by using organic polymers is not particularly high.

Metal soaps, such as aluminium stearate, are often used for façade impregnation probably because they are relatively cheap and their clearly visible water repellent effect at the surface leads to the illusion that they are particularly effective impregnating agents (Table 3.6).

Table 3.6: Effectiveness of impregnating agents

Product group	Thinner	Concentration [%]	Penetration	Reduction of water vapour diffusion [%]
Organic polymers	solvent/ water	~ 10	moderate	> 20
Metal soaps	solvent	~ 10	poor	> 20
Siliconates	water	3 to 5	moderate	10
Silanes	solvent/ water	20 to 40	very good	5 to 8
Oligosiloxanes	solvent/ water	5 to 10	very good	5 to 8
Polysiloxanes	solvent/ water	5 to 10	good	5 to 8

Organosilicon compounds (silicones)

A variety of silicone water repellents can be manufactured by various reactions of chlorosilane[87] shown in Figure 3.12.

Siliconates (A)

Siliconates are strongly alkaline solutions. They react with carbon dioxide in the air forming a silanol via an intermediate which cannot be isolated. These intermediates can react together to form a polysiloxane resin. Methyl- or propyl-potassium siliconate are the main siliconates used. Reaction with CO_2 forms potassium carbonate which deposits as a salt on the surface. Because of this formation of a white layer and the danger that the siliconate can be washed out by rain in the early stages, they are not very suitable for making façades water repellent.

Silicone resins (B)

Problems with the use of siliconates for waterproofing façades led to the use of silicone resins for this application. The relative molecular mass of the silicone resins lies between 1,000 and 3,000 and is thus very low compared with that of organic resins. Because of their relatively small molecular size, silicone resins penetrate into the pores and capillaries of building materials much better than organic resins. After evaporation of the solvent (organic solvent or water) the water repellent effect is already fully developed. However there are still reactive groups available and these react with active groups in the building material or coating components. With that the molecules are enlarged by condensation of the silicone.

Silanes (C)

Right from the early days of silicone chemistry, silanes were used to impart water repellent properties to mineral materials. They are manufactured by reaction of chlorosilanes with alcohols. Their application as water repellents for façade protection began in the 1970s. Compared with the silicone resins, the smaller molecular size means that silanes have the advantage of greater penetration when impregnating building materials.

As with siliconates, but in contrast to silicone resins, silane itself is not the effective agent. Reaction with moisture leads to intermediate silanols and subsequent condensation forms silicone resin particles. This reaction is not immediate but takes some time depending on environmental conditions. Alkylmethoxy silanes are used to obtain an acceptable reaction rate. A further undesired effect is caused by evaporation of the highly-volatile silanes.

Oligomeric alkyl alkoxysilanes (D)

Joint alcoholysis/hydrolysis of chlorosilane produces low molecular weight (oligomeric) siloxanes which still have advantages such as good penetration and the ability to be used on damp substrates but have practically no vapour pressure and therefore do not evaporate. As with silanes, crosslinking is caused by moisture. Oligomeric siloxanes have the advantages of both silicone resins and silanes. Because of their sensitivity to water, they cannot be used in conventional waterborne coatings. However, introduction of amino groups allows products to be produced which are emulsions.

The high spreading power of the organo-siloxanes allows effective impregnation of even low-absorbency substrates. It is these properties in particular that allow

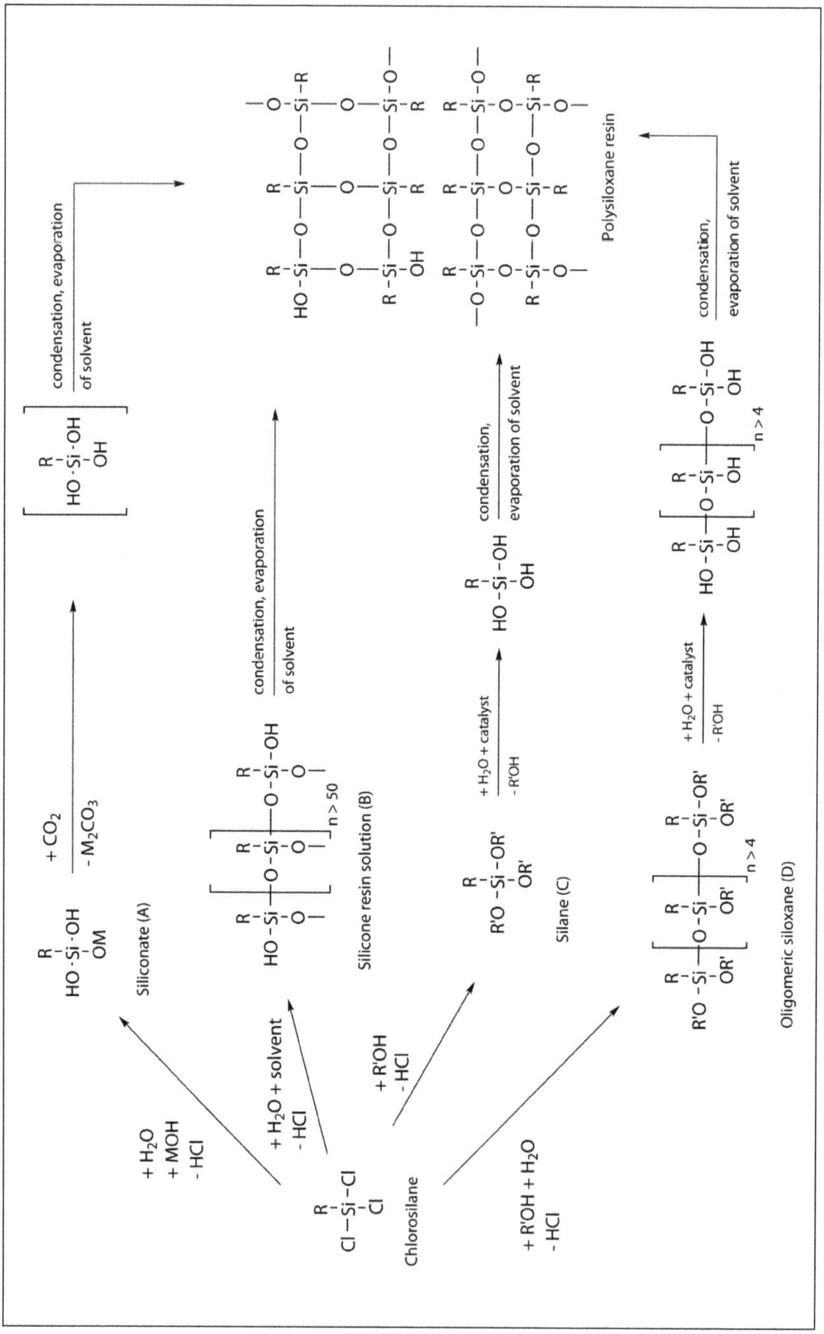

Figure 3.12: Schematic description of the chemistry for silicone water repellents

oligosiloxanes to be used in chemical horizontal treatment. This involves injecting the water repellent (usually under pressure) into the interior of the wall rather than impregnating from the surface.

As far as effectiveness is concerned, silicone resins are the best water repellents and have excellent durability and are straightforward to use.

CO_2 diffusion and rate of carbonate formation are not significantly affected by impregnation with siloxanes. These mainly depend on the quality (that is the density) of the concrete. Thus impregnation will not inhibit corrosion of the reinforcement as soon as carbonate formation has reached that point. However impregnation of reinforced concrete components is useful. It not only inhibits weathering of the carbonate layer but delays reaction of carbonates with SO_2, a reaction which leads to considerable rupture forces. Impregnation also provides effective protection against chloride corrosion which is particularly important in bridge building – where again ingress of water is the indirect cause of the damage.

Concentration of impregnating solutions

Silane-based hydrophobing materials have a low viscosity and can thus penetrate deeply into the building material; they are therefore generally not further thinned with solvents.

In contrast oligomeric alkyl alkoxysilane and silicone resins should be thinned with solvent in order to obtain sufficient penetration. Practical experience has shown that the solids content should not exceed 10 %.

3.4.2 Internal water repellency

Water uptake can also be reduced by incorporating the water repellent as a component of a coating. This internal water repellency reduces water pick-up at the coating level as soon as the film forming process has started. Among organic polymers only metal soaps and fluorocarbons can be used. However, the cost of fluorocarbon resins means that they are only used in very high-quality systems. Organo-silicon compounds are excellent for imparting this type of water repellency known as the "beading effect" (see Figure 3.13). Use of silicones reduces the susceptibility to dirt

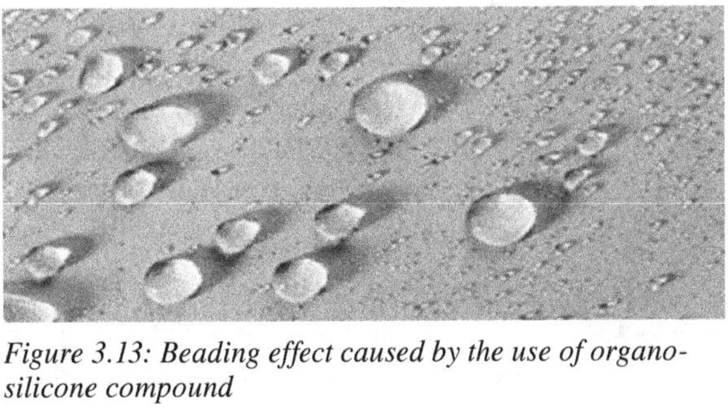

Figure 3.13: Beading effect caused by the use of organo-silicone compound

pick-up of coatings based on emulsion and silicate emulsion coatings – the water cannot penetrate the coating but is rejected directly at the surface so that the dirt is washed away. The strong water repellent effect of the silicones also causes the water to form droplets.

Water repellency depends on the orientation of the molecules in the water repellent. For example, with polyorgano-siloxanes, the siloxane dipoles are oriented towards the substrate surface while the outer layer is formed by closely packed methyl groups. Water repellency is raised by the close packing of the methyl groups; metal stearates show an analogous effect. The water repellent effect achievable with polyorgano-siloxanes is influenced by the substituents attached to the silicon atom. Silanes and siloxanes with higher alkyl groups are often more strongly water repellent than similar compounds with methyl groups. However water repellency is markedly reduced by introducing phenyl groups. Use of silanes and siloxanes with reactive groups increases the possibilities. If the performance of silicones is compared with that of paraffins it can be seen that the difference lies less in the angle of contact which can be achieved but far more in the behaviour towards the substrate. Silicones are particularly strongly attached to the substrate, spread very strongly on the surface and display their typical properties even in thin films. The porosity of the building material is little affected by silicones, meaning that permeability to air and water vapour remains unchanged. However, because of the increase in surface tension, liquid water cannot penetrate into the pores.

In practice, paraffins (saturated hydrocarbons with the general formula C_nH_{2n+2} with n in the range 25 to 50) polyethylene and polypropylene waxes are used beside polyorgano-siloxanes. These polymers have relative molecular masses between 1,000 and 4,000. They are insoluble in water or organic solvents and are used in the form of emulsions (aqueous or solvent based) of micronized material. Polyfluorocarbons (polymers of fluorinated unsaturated hydrocarbons) are also often used. The relative molecular masses of those products used for water repellent applications are between 1,000 and 5,000 and application is in the form of polymer emulsions or emulsions of micronized solids.

Figure 3.14: Contact angle

Water repellent characteristics depend on the solid/liquid contact angle between the water and the boundary surface. The contact angle (see Figure 3.14) is often measured using apparatus produced by Krüss Company of Hamburg. A micro-spray produces water droplets on the coating surface and a video camera records the contours of a liquid drop on the surface. This is digitized and displayed as a two-dimensional image on a monitor. Emulsion based architectural coatings have contact angles of about 80° whereas silicone resin coatings have contact angles of around 130°.

However exact measurement of the contact angle is very difficult. For example it is strongly affected by the surface structure of the solid and many other factors. Therefore, in practice, the contact angle is of no value as a measure of the success or failure of a water repellent but serves only to explain water repellent effects.

3.4.3 Architectural coatings

Nowadays architectural coatings have to fulfil a multiplicity of functions and correspondingly high demands are therefore made on coatings systems. As has always been the case, appearance is of prime importance. Alongside colour, brilliance and matt surfaces are important, the latter both for historic and modern buildings. At the same time the coating must be durable and over a long period the colours must not fade, there must be negligible chalking of the surfaces and the coating must not split or peel off.

The durability of coatings depends on the resistance to weathering of the components (particularly of the binders), the physical properties and chemical and mechanical compatibility with the substrate. Additionally, architectural coatings must protect the substrate particularly from ingress of moisture and deleterious substances dissolved in water. Coatings must be easy to apply and environmentally friendly and allow several repeated applications without forming stressed layers. Waterborne coatings systems range from general architectural coatings (emulsion coatings) and synthetic resin renderings, including primers, undercoats and knifing fillers, through one-pack silicate emulsion coatings based on water glass to silicate emulsion renderings and those based on a combination of silicone resin emulsion and polymer emulsion (silicone resin coatings and silicone resin renderings).

Water glass is by far the oldest binder and was introduced for this purpose around 1820. However the rapid rise of polymer emulsions caused coatings based on this to lose all importance except for certain specialist applications in conservation of monuments. Only in the 1970s with the development of stable one-pack materials in combination with suitable polymer emulsions was it possible to re-establish silicate products. All three types of waterborne coatings serve mainly as architectural coatings providing long-term protection against UV, different weathering situations with seasonal temperature changes, aggressive airborne pollutants, dirt, micro-organisms but above all ingress of moisture with all its harmful and destructive consequences.

Of all the possible coatings systems for façades, waterborne silicone resin systems come closest to satisfying the wide range of demands placed on them. They are chemically inert and dry to a stress-free microporous layer without forming a film. Because of the part inorganic, part organic nature of the binder they are permeable to water vapour and CO_2 but strongly water repellent. They adhere to all mineral substrates and also to old organic coatings, and can be re-coated many times.

Requirements for an effective architectural coating

Bearing in mind the experience of the last 30 years, the property profile for an architectural coating should fulfil the following[88]:

- low water pick-up (resistance to heavy rain)
- good water vapour diffusion
- good adhesion to mineral and non-mineral substrates
- films to have minimum stress
- excellent weather resistance
- able to be tinted using standard pigment pastes.

3.4.4 Silicate emulsion coatings and renderings

As mentioned earlier, silicate coatings are among the oldest coatings of the industrial age. Water glass is not a uniform chemical compound but a general name for glassy, stiff melts of different alkali silicates and solutions made from them. Water glasses are alkali metal salts of silicic acid. Water glass solutions are characterized by the mole ratio of silicic acid (SiO_2) and alkali oxide (Me_2O) or by weight and density.

Because of their technological advantages, silicate coatings have displaced previous architectural coatings based on lime or casein. The mid-1950s saw the development of one-pack coatings with previously unachievable storage stability as a result of the addition of a polymer emulsion. These one-pack silicate coatings currently make up the majority of the silicate coatings in use. The increasing importance of silicate emulsion as coatings for mineral substrates is due both to technical and environmental reasons. The differences between silicate emulsion coatings and numerous other coatings stem from the chemical and physical character of the water glass binder. Most coatings have organic binders but water glass is an aqueous solution of an inorganic solid. This determines the special property spectrum of silicate emulsion coatings: mineral appearance, low tendency to pick-up dirt, solvent and flame resistance, microbial resistance and the ease with which they silicify with mineral substrates. Their high permeability to water vapour and CO_2 make silicate emulsions suitable for otherwise difficult, lime-containing substrates. The mineral appearance and high gas permeability arise from the morphology of the coating. Micromorphological examination shows that the coating formed by a silicate emulsions has open pores similar to those obtained with a highly-filled interior emulsion based coatings. Silicate emulsion coatings are almost ideal for mineral building materials.

In recent years so-called sol-silicate coatings have been commercialized. These involve a combination of a pre-silificated water glass with reduced amount of alkali, and a classical water glass. Silicate emulsion coatings formulated from this have good adhesion properties even on non-silicate substrates.

However the high water vapour diffusion is associated with a high absorption of liquid water which can only be reduced by using water repellent substances to obtain systems which are totally waterproof and yet can breathe.

Investigations have shown that metal soaps do not reduce the water absorption of silicate emulsion coatings although they do have a positive influence on formation of water droplets. However this effect fades rapidly. Linear siloxanes also do not reduce water absorption characteristics probably because they do not contain

Table 3.7: Guiding formulation for a silicate emulsion coating with water-beading effect

Pos.	Components	p.b.w.	Trademark holder
1	Water	22.3	
2	"Betolin" V30	0.1	Wöllner GmbH & Co. KG
3	"Tego" Dispers 735 W	0.3	Evonik Industries AG or one of its subsidiary companies
4	"Betolin" Q 40	0.3	Wöllner GmbH & Co. KG
5	"Tego" Foamex 825	0.2	Evonik Industries AG or one of its subsidiary companies
6	"Natrosol" 250 HHR	0.3	Hercules Incorporated
7	"Kronos" 2310	10.0	Kronos Titan GmbH
8	"Omyacarb" 5	30.0	Omya AG
9	"Plastorit" 000	5.0	Imerys Talc Austria GmbH
10	"Tego" Phobe 1401	4.0	Evonik Industries AG or one of its subsidiary companies
11	"Acronal" S 559	6.0	BASF SE
12	White spirit	1.5	
13	"Betolin" K 28	20.0	Wöllner GmbH & Co. KG
Total		100.00	

any reactive groups and are therefore unable to react to form three-dimensional polysiloxanes. Although such products increase the water-beading effect, they do not lower water absorption sufficiently. Only silicone resins and oligomeric siloxanes with amino functions reduce water absorption adequately. The high effectiveness of oligomeric siloxanes stems from the amino group's high reactivity (a guide formulation is shown in Table 3.7). These products also have a very good water-beading effect. Use of highly effective silicone-based water repellents as additives in silicate emulsion coatings thus permits very effective protection against moisture. Optimal façade protection however requires a combination of external water repellency of the substrate with internal water repellency of the silicate emulsion coatings.

3.4.5 Emulsion based coatings with silicate character (SIL coatings)

Water vapour diffusion can be controlled by having a low proportion of binder and a suitable choice of filler but this leads to inadequate mechanical strength (as can

Table 3.8: Guiding formulation of a SIL coating with reduced water absorption

Pos.	Components	p.b.w.	Trademark holder
1	Water	14.7	
2	"Tylose" MHB 10.000 YP2	0.2	SE Tylose GmbH & Co. KG
3	"Tego" Dispers 715 W	0.3	Evonik Industries AG or one of its subsidiary companies
4	"Tego" Foamex 825	0.3	Evonik Industries AG or one of its subsidiary companies
5	"Kronos" 2310	12.0	Kronos Titan GmbH
6	"Omyacarb" 5	10.0	Omya AG
7	Talkum AT 1	1.5	
8	"Sikron" SF 3000	35.0	Quarzwerke GmbH
9	"Acticide" MBS	0.2	Thor GmbH
10	"Tego" Phobe 1401	2.5	Evonik Industries AG or one of its subsidiary companies
11	White spirit	0.8	
12	"Dowanol" DPnB	0.8	The Dow Chemical Co.
13	"Desavin"	0.7	Bayer AG
14	"Acronal" S 559	21.0	
Total		100.00	

be characterized by measurement of the scrub resistance) and increased chalking. Low-stress architectural coatings have been developed to overcome this. Diffusion is promoted by special pigmentation of the film. Such systems exhibit the desired property but suffer the disadvantage of increased water absorption.

Emulsion based coatings with silicate character have particularly open pores and are capillary active and water vapour permeable because they are formulated with quartz powder or other silica fillers (Table 3.8). The resultant capillary water absorption can be reduced by addition of amino-functional siloxanes without affecting the water vapour diffusion.

3.4.6 Siloxane architectural coatings with strong water-beading effect

These coatings consist of a polymer emulsion, a water repellent additive, fillers, titanium dioxide and, if necessary, other standard additives. It might be expected that these coatings would involve a silicone resin (see following section) but the special point about this formulation is the use of functional polysiloxanes instead. Like the SIL coatings just described but unlike the formulation of silicone resin coatings, quartz powder is used although with a considerably smaller particle size distribution (maximum 13 µm) and a larger amount of titanium dioxide. The polymer emulsion used should contain as few water soluble components as pos-

Table 3.9: Guiding formulation of a siloxane façade coating with strongwater-beading effect

Pos.	Components	p.b.w.	Trademark holder
1	Water	29.7	
2	"Tego" Foamex 825	0.1	Evonik Industries AG or one of its subsidiary companies
3	"Surfynol" E 104	0.25	Air Products and Chemicals, INC.
4	"Walocel" XM 6000 PV	0.3	Dow Wolff Cellulosics GmbH
5	"Acticide" MBS	0.1	Thor GmbH
6	"Kronos" 2044	20.0	Kronos Titan GmbH
7	"Sibelite" M 3000 or "Calcimatt"	32.0	SCR-Sibelco naamloze vennootschap/ Omya AG
8	Ammonia 25 %	0.15	
9	"Tego" Phobe 1505	2.4	Evonik Industries AG or one of its subsidiary companies
10	"Acronal" S 790	15.0	BASF SE
Total		100.00	

sible, i.e., it should be low in emulsifiers. This is, of course, equally true for the additives. In addition to its low water absorption, the coating is also characterized by very good water vapour diffusion. The unique micro-pimple texture obtained by this formulation leads to a very strong beading effect (Table 3.9). At first it was thought that the beading effect would result in lower tendency to pick-up dirt but it was later established that water repellency decreases relatively rapidly as the length of exposure to weathering increases and that this can result in increased dirt pick-up. Possible reasons for this will be dealt with in the next section.

3.4.7 Silicone resin coatings and renderings

As already mentioned, silicate coatings, which have a porous structure, were often used as architectural coatings. These materials permit high water absorption by the façade but also allow the façade to dry out by diffusion of water vapour. With the introduction of emulsion-based architectural coatings, pore-free systems were used enabling water absorption to be considerably reduced. However these coatings formulated below the critical pigment volume concentration (CPVC) exhibit only low water vapour permeability. This property profile is very good only if the substrate is a dry wall. However, as previously discussed, building errors often mean that a wall does not remain dry. Damage to the roof or a defective horizontal barrier above the foundation can allow water to enter. As the water cannot escape quickly enough through pore-free coatings, building damage such as spalling can occur. It is true that water vapour diffusion can be improved by reducing the amount of binder and targeted choice of fillers (SIL coatings) in the formulation, however this leads to higher water absorption due to the high absorbency of capillaries in the "above-critical" PVC systems. Formulation of coatings (silicone resin coatings), in which the required application profile (Künzel, see Chapter 5.1) is largely fulfilled, involves using a combination of acrylic emulsions with silicone resin emulsions and a high degree of filling.

In silicone resin coatings the water pick-up is reduced without the pores and capillaries being closed. The silicone resin lines the surface of the capillaries with a water repellent film which inhibits wetting by water. There is thus irreversible absorption of the silicone resin onto the fillers contained in the silicone resin coating.

The silicone resin molecules orientate themselves because of their mobility so that the methyl groups of the resin are on the outside and this again leads to very strong water repellency.

The long life of silicone resin coatings, especially in respect of weathering and water repellency, can be traced back to the fillers and the formation of silicone networks. The real difference between silicone resin coatings and exterior emulsion based coatings is the very high degree of pigmentation and the use of a silicone resin emulsion alongside the polymer emulsion.

The reason for combination with a polymer emulsion is primarily that silicone resins have too little pigment binding power to ensure adequate fixing of the pigments and fillers. The pigment binding power of a polymer emulsion is the ability to bind together as many pigments and fillers as possible. Determination and evaluation of pigment binding power requires a clear understanding of the definition and meaning of the critical pigment volume concentration, CPVC (see Chapter 5.4). The CPVC is the pigment volume concentration (PVC) at which the quantity of binder is just sufficient to completely wet pigments and fillers and fill the interstices, the pigments and fillers being as closely packed as possible. The CPVC could also be described as the point at which the system changes from being a polymer (binder) dominated structure into a pigment/filler dominated structure.

The term "critical" arises from the fact that if the CPVC is exceeded most film properties alter almost stepwise. For example, opacity, porosity and permeability increase strongly, above the CPVC and the ultimate tensile stress and scrub resistance decrease sharply. Above the CPVC, film porosity increases, that is the s_d-value becomes much smaller, i.e., the water vapour permeability of the coating material is increased.

This desired porosity of the film results, however, in higher water absorption so that, in the final analysis, both opposing physical properties need to be taken into account during formulation. The coefficient of water absorption gets smaller as the amount of silicone resin emulsion increases and the water resistance improves. At the same time the water vapour permeability deteriorates. This is because the pores necessary for water vapour permeability not only become lined but gradually fill up. At a 1:1 ratio of polymer emulsion to silicone resin emulsion, the water vapour permeability and the coefficient of water absorption are both the level required. The pore walls have absorbed silicone resin thus ensuring water repellency. Despite this there is still enough pore volume for sufficient water vapour permeability. The extremely high CPVC of 65 % also guarantees an optimal price/performance ratio. The difference between the chosen PVC and the CPVC should be 15 PVC units.

The type and quantity of silicone resin emulsion to be used must be determined carefully remembering that at every addition the pore volume and with it the water vapour permeability decreases. The pores must not become filled with the silicone resin, only lined with it; that is the added silicone resin volume must always be less than the pore volume. To obtain good resistance to driving rain, methyl/phenyl-silicone resins are preferred to methyl-silicone resins (Chapter 2.1).

In other words, silicone resin coatings and renderings achieve the demand for low diffusion resistance and low water absorption by achieving a water repellent porosity.

New waterborne methyl-silicone resins enable the manufacture of silicone resin coatings with smaller silicone resin content.

This makes it possible to achieve s_d-values in Class 1 and w-values in Class 3 to DIN EN 1062-1 with 4 % of a 50 % silicone resin emulsion.

Tendency of waterborne silicone resin coatings to pick-up dirt

From experience it is known that silicone resin architectural coatings have a very low tendency to pick-up dirt, especially compared with emulsion based coatings which have a low PVC[89]. This probably stems from the different wetting behaviour of the resulting coating films. These dirt particles can accumulate on the façade especially in regions where rainwater contains concentrated matter such as soot. Dry dirt particles carried to the façade on the wind seem to be less of a problem than organic dirt. Open-air weathering for 3 years vertically facing south shows that for silicone resin coatings and renderings based on polymer emulsions

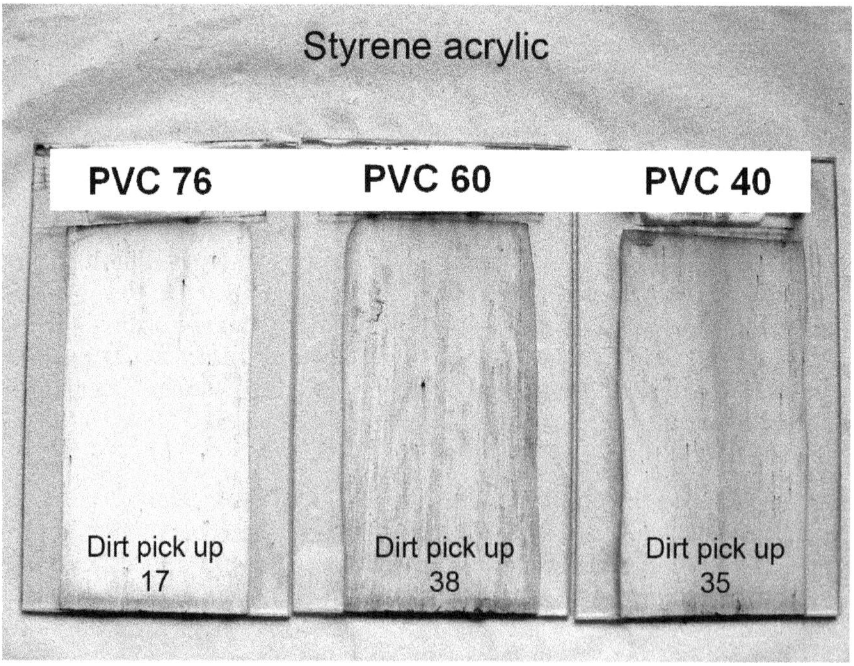

Figure 3.15: Dirt pick-up of waterborne silicone resin coaating as a function of the PVC

with a glass transition temperature > 10 °C there is no significant difference in dirt pick-up or chalking properties between pure acrylates and acrylate/styrene copolymers (all other materials remaining the same). In formulating solvent-free silicone resin systems however, the choice of starting materials has a measurable effect on the result. Thus, solvent-free silicone resin emulsions have a slightly lower, but not very significant, dirt pick-up than those containing solvents.

However the most important influence appears to be the PVC (Figure 3.15). As it takes at least 6 months to even begin to evaluate the tendency to uptake dirt by an architectural coating, Evonik (Business Line Coating Additives) has developed an accelerated test to determine dirt uptake of façade coatings. In this test, glass panels are coated with the architectural coating using a doctor blade with a slit height 300 µm. The plates are then dried for 24 hours at 50 °C in a convection oven.

The coated glass panel is placed at an angle of 70° in a dirt machine which is a converted industrial dishwasher (Figure 3.16).

The panels are washed with water at 30 °C for 4 minutes, followed by contamination with a 0.1 % dirt solution, consisting of tar, soot, a standard inorganic dirt, water and solvent, at 30 °C for 1 minute. The contaminated panels are dried for 10 minutes at 50 °C. The whole cycle is repeated 10 times. The difference in L-value, before and after contamination is considered as the dirt pick-up of the sample.

This accelerated test correlates with open-air weathering in Jakarta/Indonesia where the dirt pick-up is predominantly organic. However, there is no correlation between this accelerated test and open-air weathering in Europe. Comparison

Figure 3.16: Accelerated test to determine dirt pick-up of façade coatings

of open air weathering tests within Europe showed that the local pollution and climate have a significant effect on the results.

Correlation between the sites has been limited. The results from individual sites, as well as from the dirt machine need to be considered as unrelated. It is therefore difficult to generalize the results.

The results of the outdoor tests confirmed that the PVC has the greatest influence irrespective of the polymer emulsion used. Only polymer emulsions with glass transition temperatures below 10 °C pick-up more dirt. The reason that solvent-free silicone resins have a somewhat lower dirt pick-up than those containing solvent may be due to the fact that water-soluble components are more strongly

Table 3.10: Guiding formulation for a silicone resin coating based on nanohybrid emulsion

Pos.	Components	p.b.w.	Trademark holder
1	Water	25.2	
2	"Tylose" MH 30000 YG8	0.2	SE Tylose GmbH & Co. KG
3	"AMP-90"	0.3	The Dow Chemical Company (Dow) or an affiliated company of Dow
4	"Tego" Dispers 715 W	0.4	Evonik Industries AG or one of its subsidiary companies
5	"Agitan" 265	0.3	MÜNZING CHEMIE GmbH
6	"Acticide" MBS 50:50 Methyl-/ Benzisothiazolinone 1:1	0.1	Thor GmbH
7	"Acticide" F (N)	0.1	Thor GmbH
8	"Kronos" 2056	15.0	Kronos Titan GmbH
9	"Finntalc" M 30	5.0	Mondo Minerals BV
10	"Dorkafill" H	10.0	Gebrüder Dorfner GmbH & Co. Kaolin- und Kristallquarzsand-Werke KG
11	"Omyacarb" 5 GU	10.0	Omya AG
12	"Omyacarb" 15 GU	10.0	Omya AG
13	Butyl diglycole acetate	1.2	
14	"Tafigel" PUR 40	0.2	Münzing Chemie GmbH
15	"Tego" Phobe 1650	6.0	Evonik Industries AG or one of its subsidiary companies
16	"Mowilith" LDM 7717	8.0	Celanese Emulsions GmbH
17	"Mowilith" Nano 9420	8.0	Celanese Emulsions GmbH
Total mill base		100.00	

dissolved out of the dry film. It could be suggested that the difference in dirt uptake is related to a certain extent to different chalking behaviour. Stronger chalking should result in lower dirt uptake as the chalking out, i.e., degradation of the top layer, also removes dirt on the surface. If this effect exists, it will only be evident in open-air weathering and not in the accelerated test.

Silicone resin coatings involving combinations of silicone resins and polymer emulsions based on styrene acrylates and pure acrylates already have a low dirt uptake. This can be further reduced using a new generation of emulsions made from colloidal silica-latex polymer nanocomposites (hybrid polymers). Experience indicates that a triple combination of an acrylate emulsion, a nanohybrid emulsion and a silicone resin emulsion as in the formulation below is particularly effective (Table 3.10).

3.4.8 Below-critical PVC formulated exterior coatings

Especially in Southeast Asia, the use of silicone resin coatings continues to be unpopular. The reason for this is the predominant use of silk-finish coatings which are below-critical formulated architectural coatings. However there are also doubts as to whether above-critical formulated exterior coatings can cope with climatic conditions there. All attempts to formulate below-critical silk-finish exterior coatings using silicone resins have failed. Consequently it has been necessary to pursue other options.

A new technology has finally made it possible to formulate exterior coatings which achieve the s_d- and w-values demanded by DIN-EN 1062-1. This new method involves an extendable core-shell acrylate emulsion. Transport of water vapour is accelerated by locally concentrated polar groups without creating pores in the film after coalescing.

By replacing up to 20 % w/w of the polymer emulsion used in the particular formulation, the s_d-value can be reduced by around 50 to 80 %. The most marked effect is achieved with PVC between 20 to 40 %.

The frequently desired water beading effect can be achieved by targeted addition of polysiloxanes (Table 3.11).

3.4.9 Photocatalytic architectural coatings

Dirt pick-up by façades has been investigated for many years. However it is a complex phenomenon and requires care in optimizing the formulation for all the requirements to be satisfied.

It is known that atmospheric organic compounds which are toxic and difficult to decompose are fully oxidized in the presence of hydroxyl radicals. It is thus possible to alter very different materials by applying extremely thin, transparent TiO_2 films so that oily substances are decomposed under the influence of solar UVA radiation, thus markedly reducing the ability of dirt particles to adhere to the surface.

The Fraunhofer Institute has discovered that photocatalytic applications require amorphous TiO_2 films with large surface area[90].

Table 3.11: Guiding formulation for a silk-finish emulsion coating with low water absorption and high water vapour transmission

1. Mill base			
Pos.	Components	p.b.w	Trademark holder
1	Water	21.0	
2	"Tylose" 6000 YP 2	0.2	SE Tylose GmbH & Co. KG
3	"Tego" Foamex 855	0.1	Evonik Industries AG or one of its subsidiary companies
4	"Calgon" N (solution, 10 %)	0.2	Reckitt Benckiser N.V.
5	"Tego" Dispers 755 W	0.4	Evonik Industries AG or one of its subsidiary companies
6	"Kronos" 2190	24.2	Kronos Titan GmbH
7	NaOH (solution, 10 %)*	0.6	
Total mill base		46.7	
2. Let down			
Pos.	Components	p.b.w	Trademark holder
8	"Mowilith" LDM 7714	38.8	Celanese Emulsions GmbH
9	"Tegovapro"	12.1	Evonik Industries AG or one of its subsidiary companies
10	"Acrysol" RM 5000	-	Rohm and Haas Company
11	"Texanol"	2.4	Eastman Chemical B.V.
Total let down		53.3	
Total		100.0	
PVC (%)		20	
Manufacture			

* Neutralization with ammonia or organic amines need to be avoided, which may cause a coagulation of the emulsion

Research is now taking place to use this principle in architectural coatings by incorporating photocatalytically-active TiO_2 in a binder matrix so that the effects described can occur on a normal façade. UV irradiation of a semi-conductor such as titanium dioxide or silicon dioxide in the form of nanocrystalline particles converts water and oxygen to reactive hydroxyl radicals. Oxides of nitrogen (NOx) are converted to nitric acid which reacts with calcium carbonate in the formulation. Oxides of nitrogen are chemically bound and water and carbon dioxide liberated[91]. Effectively a combustion reaction occurs, liberating water, CO_2 and other innocuous products, see Figure 3.17.

Table 3.12: Guiding formulation of a waterborne façade coating with photocatalytic effect

Pos.	Components	p.b.w.	Trademark holder
1	Water	8.5	
2	Photoactive TiO_2	20.0	
3	"Tego" Dispers 715 W	0.2	Evonik Industries AG or one of its subsidiary companies
4	"Calgon" N	0.2	Reckitt Benckiser N.V.
5	"Acticide" MBS	0.2	Thor GmbH
6	"Acticide" MKA	1.0	Thor GmbH
7	"Walocel" XM 20000 PV	0.3	Dow Wolff Cellulosics GmbH
8	"Lusolvan" FBH	1.0	
9	"Kronos" 2190	12.0	Kronos Titan GmbH
10	"Omyacarb" 2 GU	12.0	Omya AG
11	"Omyacarb" 5 GU	10.0	Omya AG
12	"Socal" P 2	4.0	Solvay GmbH
13	"Optimat" 2550	1.2	Imerys Minerals Ltd.
14	"Tego" Phobe 1650	11.3	Evonik Industries AG or one of its subsidiary companies
15	NaOH	0.1	
16	"Acronal" 290 D	17.0	BASF SE
17	"Tego" ViscoPlus 3030	0.5	Evonik Industries AG or one of its subsidiary companies
18	"Tego" Foamex 855	0.5	Evonik Industries AG or one of its subsidiary companies
Total		100.00	

As the usual binders for architectural coatings (e.g. pure acrylates and styrene acrylates) are decomposed by catalysis relatively quickly, the question of a suitable binder is of importance. A possible starting point for investigations to resolve this is the use of water glass or silicone resin emulsions Table 3.12).

$$2\ HNO_3 + CaCO_3 \longrightarrow Ca(NO_3)_2 + H_2O + CO_2$$

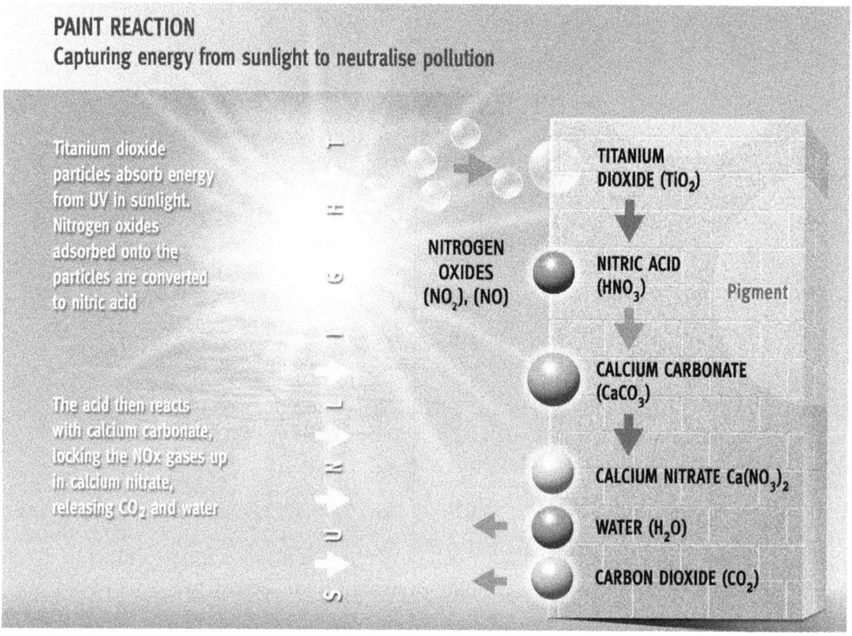

Figure 3.17: Capturing energy from sunlight to neutralize pollution

4 Outlook

World-wide consumption of pure silicone resins and silicone combination resins (see Chapters 2.1 and 2.2) is expected to increase from 86 thousand metric tons in 2012 to about 103 thousand metric tons in 2018. Silicone resins are supplied as 100 % resins, solvent or solvent-free systems, emulsions or in powder form.

The specific properties of silicones – heat resistance, hardness, flexibility and water repellency – have made them unique raw materials for coatings. Without heat resistant silicone resins, barbeques and car silencers could not be finished with a durable coating and saucepans, frying pans and other consumer goods which are exposed to high temperatures would not keep their good looks for very long. Because of their excellent temperature resistance, silicone resins with reactive groups are used to modify polyester, alkyd, epoxy and acrylic resin enamels used in coil coatings, as well as in anti-corrosion and maintenance coating applications.

The alkoxy-silyl modification of organic resins and isocyanate crosslinkers make automotive finishes scratch-proof and architectural coatings based on silicone resins more durable than those used previously.

Dirt repellency, microbial resistance and durability of exterior façade coatings are still the most frequent problems. Another challenge for façade coatings is easy removal of graffiti.

A further, increasingly important, aspect is the reduction of VOCs which can be achieved by the polysiloxane presence in silicone combination resins. With silicone combination resins, very high solids content in the binder and final coating formulation are possible at low viscosities, a situation impossible to achieve with pure organic binders. This is because the viscosity is reduced by the polysiloxane component in silicone combination resins. In view of the increasingly stringent regulations concerning reduction of VOCs, this aspect will further reinforce the interest shown in recent years in silicone combination resins.

W. Heilen, S. Herrwerth: Silicone Resins and their Combinations

ISBN: 978-3-86630-828-2

5 Glossary

5.1 Façade protection theory according to Künzel

Despite the water repellent effect already described, coatings are not completely impermeable to water or water vapour. For façade protection it is also necessary that a mineral building material has a certain ability to "breathe".

In Künzel's façade protection theory, two properties of the coating play an important role both for protection and breathing. These are the water absorption capacity of the building material and the water vapour permeability of the coating. Because of the protective effect of the coating, transport of water vapour from the outside to the interior is very slow. The process is one of diffusion and the rate depends strongly on the concentration gradient between the substrate and the surrounding air.

Façades designed according to these guidelines lose more water in the drying phase than they absorb when wetted by rain.

Water repellent coatings satisfy both the requirements of gas permeability and capillary water absorption and are therefore the highest quality products on the market with an almost ideal performance.

5.2 s_d-value

According to Künzel, a quantitative relationship can be set up between the water repellency of a coating and its permeability to water vapour. In this, two parameters, the s_d-value and the w_{24}-value, play a decisive role. The s_d-value is the thickness (in m) of a hypothetical, static air layer which would have the same resistance to water vapour diffusion as the coating or building material. It is obtained by measuring water vapour diffusion to EN ISO 1062-1. For a coating film this depends on the coating thickness and the degree of pigmentation that is the PVC. s_d-values for coatings range from < 0.1 m for water-permeable films (e.g. silicate coatings) to about 2 m for coatings with very little microporosity (e.g. solvent containing acrylic resin coatings).

W. Heilen, S. Herrwerth: Silicone Resins and their Combinations

ISBN: 978-3-86630-828-2

A European standard for classification of water vapour permeability has been developed (EN ISO 1062-1), see Table 5.1.

Table 5.1: Classification of EN ISO 1062-1

Water vapour transmission rate, s_d-value (EN ISO 1062-1)		
Class	Water vapour diffusion	s_d-value
I	high	< 0.14
II	medium	0.14 to 1.40
III	low	> 1.40

Determination of water vapour permeability (wet-cup method)

Water vapour permeability is determined gravimetrically using an inert substrate: 60 mm porous frits are used. The coating is distributed evenly over the frit using a flat brush. A second coat is applied after drying. The dry film thickness should be 150 µm. Drying takes place overnight under standard climatic conditions (23 °C/50 % relative humidity (EN ISO 7783)) followed by 24 hours drying at 50 °C.

A weighing glass with a diameter of 61 mm and height of 30 mm is filled with 20 ml of distilled water. To avoid wetting the underside of the frit, which could

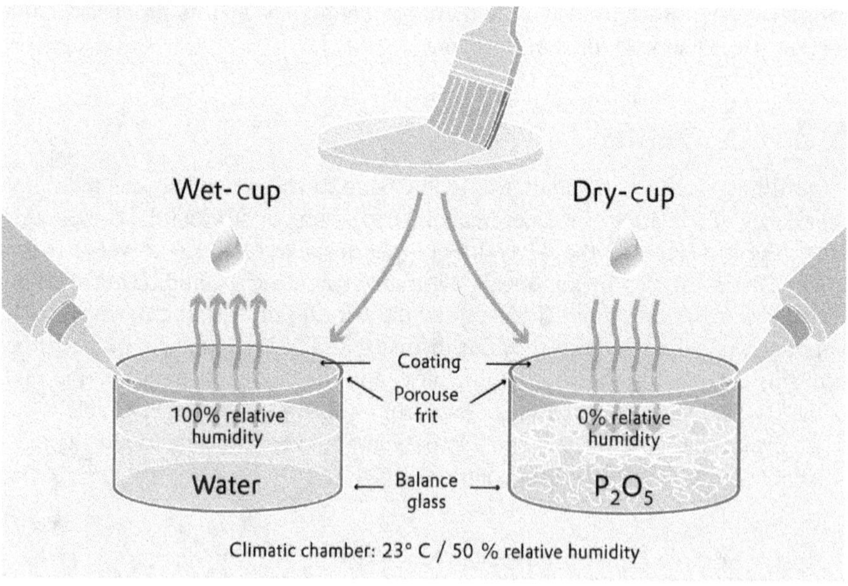

Figure 5.1: Measuring water vapour permeability (wet-cup and dry-cup method)

cause errors, a sponge is put into the weighing glass. The coated frit is then placed in position and a hot melt adhesive used to seal between the frit and the vessel.

The prepared weighing glass is weighed on an analytical balance (to ± 0.1 mg) and then placed in a climatic chamber at standard conditions 23/50 for 5 days. The amount of water vapour diffused is then obtained by a further weighing.

Comparison of the wet-cup and dry-cup methods for determining water vapour diffusion

In the wet-cup method the diffusion gradient is from 100 % relative humidity in the glass to 50 % relative humidity in the climatic chamber. In the dry-cup method the direction of diffusion is reversed, 0 % relative humidity in the glass to 50 % in the climatic chamber. The amount of water diffused in the wet-cup method is 10 times that in the dry-cup method providing greater accuracy. The wet-cup method is thus preferred in practice.

5.3 w-value

The w-value is a measure of the water absorption of the building material via its capillaries. The w-value is calculated from the weight of water absorbed (in kg) divided by the product of the surface area (in m^2) and the square root of the time. The values of the parameter are shown in Table 5.2.

Measuring capillary water absorption

a) Preparation of the substrate

The substrate is of calcareous sandstone with an area of 115 • 70 mm = 0.008 m^2 and of thickness 20 mm.

The cut stones are thoroughly scrubbed under water and then dried for 24 hours at 50 °C.

b) Coating the substrates

The individual layers of the coating system are then poured onto the stone and spread with a flat brush, coating the side walls as well as the main surface (400 ml/m^2). Care is taken to ensure that all pores are closed.

Table 5.2: Classification of w-values

Water permeability classes, w-value (EN ISO 1062-1)		
Class	Water uptake	w-value
III	low	< 0.1
II	medium	0.1 to 0.5
I	high	> 0.5

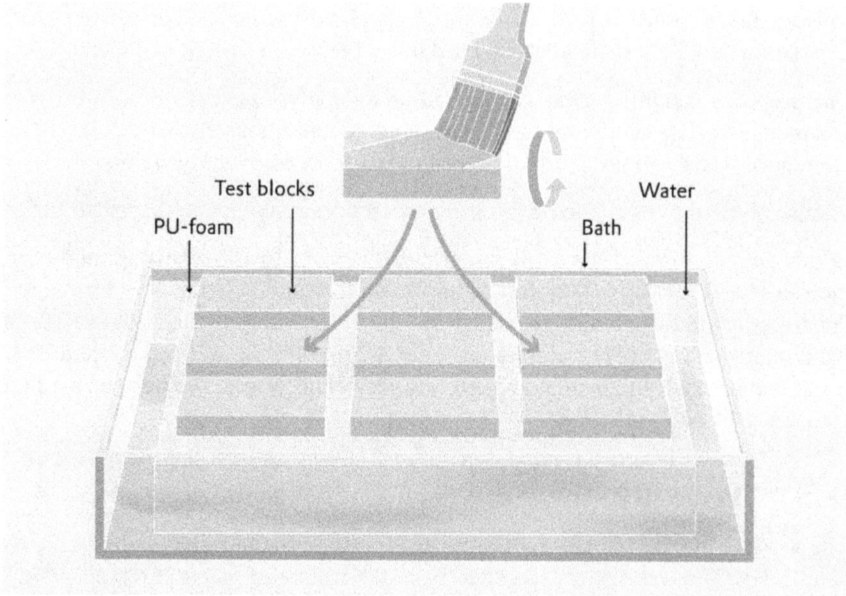

Figure 5.2: Measuring absorption of water

c) Exposure to water

The prepared sample is exposed to water in a dish lined with foam plastic. The water must reach the underside of the stone and from time to time the water level is checked and topped up as necessary. After 24 hours the stone is removed from the water bath, laid on paper and patted dry. It is then weighed and the w_{24}-value calculated.

DIN EN 1062-3 specifies that the stone should be exposed to water three times to wash out water-soluble components and a fourth exposure made to calculate the w_{24}-value.

The w_{24}-value is defined as kg water/(m^2 • (24 h)1/2). For example, if the water uptake is 0.5 kg/m^2 after 24 hours exposure to water:

$w_{24} = 0.5/\sqrt{24} = 0.5/4.9 \approx 0.1$

d) Comments on the method

The accuracy of the w_{24}-value increases with the thickness of the coating as absorption by the substrate becomes less important. As long as the barrier effect of the coating remains intact, the w_{24}-value is independent of coating thickness.

Protective systems for mineral building materials involving coating with emulsion or silicate emulsion coatings, must have s_d- and w_{24}-values as small as possible

5.4 Definitions of PVC and CPVC

PVC – Pigment volume concentration

Coatings formulations are commonly specified either in terms of weight or volume. Examples of weight-related parameters are parts by weight, weight percentage and the pigment/binder ratio, the so-called pigmentation. The latter is the ratio by weight of pigment plus filler to (solid) binder.

The main constituents of the solid coating film are pigments, fillers and binders. The effects of these constituents are in proportion to their volume. The most important volume-based formulation parameter is the pigment volume concentration, usually known by the abbreviation PVC.

The PVC is a calculated quantity: the ratio of the pigment volume to the total volume of the coating film expressed as a percentage. The PVC is determined by the formulation. It has become accepted that the term "pigment volume" is the sum of pigment and filler volumes. Consequently, a more correct term would be "total PVC" which is then defined as:

$$\text{Total-PVC} = \frac{\text{P-Volume} + \text{F-Volume}}{\text{P-Volume} + \text{F-Volume} + \text{B-Volume}} * 100\ [\%]$$

where:

P = Pigment

F = Filler

B = Binder (solid)

The volumes needed to calculate the PVC are obtained from the masses (in g or kg) and the densities of the components. In the case of binders it is the density of the pure binder (also called the solid binder) which must be used:

$$\text{Volume} = \frac{\text{Mass}}{\text{Density}}\quad [\text{m}] \text{ or } [\text{l}]$$

The total PVC of common coatings varies over a wide range; for glossy emulsion coatings the PVC lies in the range 17 to 23 %. The highest total PVCs are found with matt emulsion coatings where they are in the range 40 to 85 %.

CPVC – Critical pigment volume concentration

The CPVC is usually defined as the PVC at which the binder is just sufficient to fully wet the pigments and fillers and fill the interstices when the pigments and fillers are as closely packed as possible.

In the case of emulsion based coatings the same definition is used but it should be noted that the binder is in the form of a polymer emulsion. At the CPVC of a emulsion based coating, pigments, fillers and polymer particles are as closely packed as possible. It is true that the polymer particles are stuck together and deformed by coalescence but in their microscopic structure they are still particles.

Values of the PVC below the CPVC are spoken of as being in the "below critical" region where there is an excess of binder: pigments, fillers and interstices are completely surrounded or filled by binder but pigments and fillers are not in their closest-packed state.

The film surface is covered with a layer of fused polymer particles. This produces a gloss which depends on the fineness of the pigment, filler and polymers particles and decreases as the PVC increases up to the CPVC.

Values of the PVC above the CPVC are said to lie in the "above critical" region. There is a deficit of binder. Pigments and fillers are not completely surrounded and interstices are not completely filled. There remain spaces which are filled with air and the film is porous. The deficiency of binder means that pigments and fillers come into direct contact with air so that a dry-hiding effect is produced and there is no gloss. At even higher above-critical PVCs a gloss known as a sheen can be produced.

The CPVC lies at the interface between the below- and above-critical regions. The term "critical" is used because all film properties change markedly when crossing the CPVC.

The CPVC, or more correctly the position of the CPVC, is strictly speaking not a material constant. Although it is influenced to some extent by the type of raw materials, more importantly, it depends on their combination which, in turn, depends on the formulation.

6 Analysis of silicone polymers

6.1 NMR spectroscopy

Nuclear Magnetic Resonance Spectroscopy (NMR) is particularly suitable for elucidating the structure of silicone polymers. It uses the magnetic properties of those atomic nuclei which have a nuclear spin. Nuclear resonance instruments are in principle radio instruments linked with a transmitter and a computer. These basic units are integrated in modern NMR instruments.

The method was developed in the nineteen fifties by M. Purcell (Harvard University) who, together with F. Block was awarded the Nobel Prize in 1952 in recognition.

The method was used by E. G. Rochow (developer of the Rochow Synthesis, c.f. Chapter 1.1) to investigate the structure of methyl-silicones. Even at that time, it appeared particularly suitable since the most important isotopes of silicon, oxygen and carbon are so-called g.g. (gerade-gerade, i.e. even-even) isotopes, which have magnetic moments which are zero. The only magnetic atoms in this investigation were the hydrogen atoms of the methyl group enabling the motion of these nuclei to be observed without interference by other elements[92].

Nowadays, NMR spectroscopy (^{1}H, ^{13}C and ^{29}Si) is considered the most suitable method for the structural analysis of polymeric silicone compounds. The technique permits both quantitative and qualitative measurements.

In a magnetic field, the nuclei behave like magnets with a nuclear spin quantum number of 1/2 so that only two orientations of the axis of magnetic moment to the external field are possible.

The resonance frequency of the nuclear species is influenced by the nuclear environment. The nucleus is shielded by its environment and the magnetic field, as seen by the nucleus, is weakened, i.e., the applied field strength at a given frequency must be increased to compensate for the shielding and bring the nucleus into resonance.

W. Heilen, S. Herrwerth: Silicone Resins and their Combinations

ISBN: 978-3-86630-828-2

High resolution spectroscopy of, for example, ^{29}Si requires particularly constant, uniform magnetic fields since resonance signals must often be recorded over several hours (3 to 10 h). Nowadays, only NMR instruments equipped with electro- and super-conducting magnets are used. Current NMR instruments equipped with super-conducting magnets are operated at magnetic field strengths up to 11.5 Tesla, corresponding to a proton resonance frequency of up to 500 MHz. The magnetic field is generated via a coil made of a special Niobium alloy, the wires of which are embedded in a copper matrix.

The NMR spectrometer produces the required frequencies and transmits them to a measuring head in the magnet. The NMR signal received is amplified and sent to the computer for processing[93].

All required frequencies are derived phase-locked from a master quartz crystal.

There are three channels:

- measurement channel
- lock channel
- decoupling channel.

The exciting frequency is synthesized from the master quartz and fed via a pulse amplifier to the transmitting coil in the measuring head at a specified repetition rate. The measuring head carries both the transmitting and receiving coils for the various channels. The NMR tube containing the sample is spun.

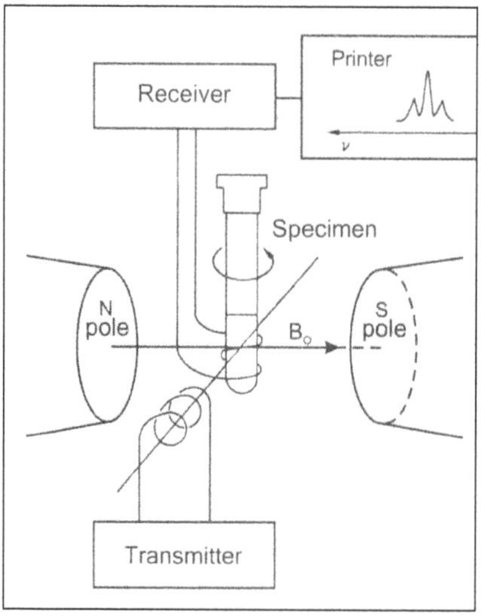

Figure 6.1: Layout of a nuclear magnetic resonance spectrometer

An oscillator produces a higher frequency and feeds this to a mixer. The NMR signal from the transmit/receive coils (measuring channel) is also amplified and passed to the mixer.

After low frequency amplification, the NMR signal is passed to a computer via an analogue to digital converter (ADC) and converted to a traditional NMR spectrum by the mathematical process of Fourier transformation. The NMR spectrum is then printed.

The locking channel is similar to the measuring channel with the difference that the demodulated signal is not fed to the computer or printer but used to fine tune the magnetic field to ensure resonance conditions.

In contrast to the two high-frequency channels just described, the decoupling channel has no receiver of its own. It serves to decouple all the protons in a specimen at the same time[94] so that the spectrum obtained is much simpler and easier to interpret.

The position of the signals in the case of ^{1}H und ^{29}Si is measured relative to a tetramethyl silane (TMS) reference and this defines the chemical shift. This chemical shift does not depend on the magnetic field or the measurement frequency and is quoted in ppm.

The fine structure of the signals, also termed multiplets, arises from the interaction of neighbouring nuclei and is quantified in terms of coupling constants.

^{1}H-NMR is particularly suitable for polymers containing tetra-coordinate silicone and permits rapid specific detection of polymeric compounds. Although standards are less important for identification, the compounds being investigated must of course contain protons. For quantitative analysis, an internal standard in the form of a suitable solvent must be added to the sample. The choice of suitable solvent is often difficult. Its resonance must be such that no interference with the sample signals occurs and all compounds which are being investigated should be soluble in it.

In the case of silicone resins, a ^{1}H-NMR spectrum provides a wide range of information on dimethylsiloxy units in the polymer and the methylsiloxy and ethoxy content[95–96].

Typical groups which can be detected with ^{1}H-NMR are shown in Table 6.1.

Table 6.1: Typical groups and their chemical shift detected with ^{1}H-NMR

Group	Chemical shift, ppm
$\equiv Si-CH_3$	0.0
$\equiv Si-CH_2-$	0.5
$\equiv Si-OCH_3$	3.5
$\equiv Si-H$	4.7
$\equiv Si-CH=CH_2$	5.8 to 6.2
$\equiv Si-C_6H_5$	7.3 (intensity 3) 7.6 (intensity 2)

^{29}Si-NMR spectroscopy (Figure 6.2) is used to

- quantify the M, D, T and Q units in the polymer
- measure the number and type of neighbouring atoms
- determine the relationship between functional and terminal groups in the polymer.

The bond angles and conformations can be inferred from the coupling constants.

Table 6.2: Typical groups and their chemical shift in analysed with ^{29}Si-NMR

Group		Chemical shift, ppm
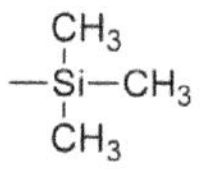	(M)	+ 8
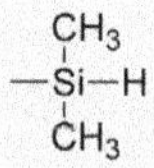		- 5
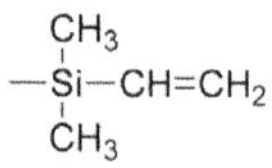		- 5
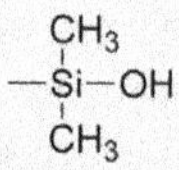		- 10
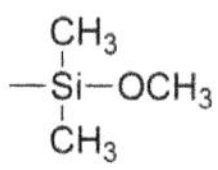		- 12
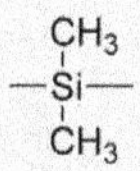	(D)	- 20
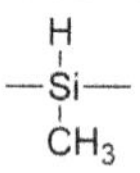		- 35
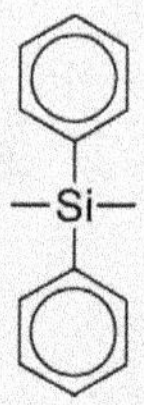		- 45
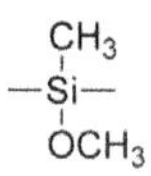		- 56
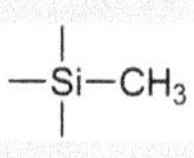	(T)	- 65
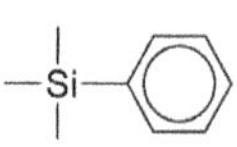		- 78
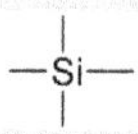	(Q)	- 100

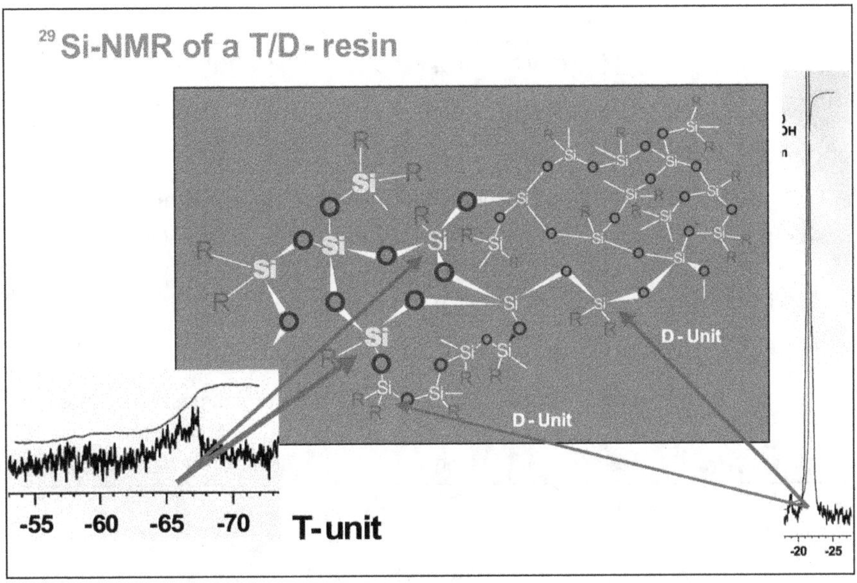

Figure 6.2: ^{29}Si-NMR spectroscopy of a T/D resin

Table 6.2 shows typical groups which can be analysed with ^{29}Si-NMR.

6.2 IR spectroscopy

IR spectroscopy is useful for establishing the chemical composition. The infrared siloxane or other absorption bands may be used to determine the silicones or silicates in a mixture. Silicones are identified and determined by mean of their IR spectra; the $\equiv SiCH_3$ and $\equiv Si(CH_3)_2$ groups absorb at 1259 and 800 cm-1, respectively; and Si – O – Si at 1010 to 1110 cm^{-1}. The ratio of methyl to phenyl groups is determined by measuring the intensity of $\equiv SiCH_3$ and $\equiv SiC_6H_5$ bans at 1263 and 1435 cm-1, respectively. Si – H is determined by IR absorption at 2100 to 2250 cm^{-1}. The methyltriacetoxysilane concentration in continuously compounded RTV sealants is monitored online by automated laser-IR spectroscopy[97].

The most important IR bands for silicone polymers are summarised in Table 6.3.

6.3 Wet analysis

Special groups are determined by specific chemical reactions. For example, chlorosilanes are hydrolyzed, and the halogen is determined by titration with alkali or silver nitrate. Other types of halogen substitution may require more drastic methods of decomposition. Silicon hydride (Si – H) is assayed by determination of

Table 6.3: Important IR bands for silicone polymers

Group	Wave number cm^{-1}	Intensity of band *
$\equiv SiOH$	3685 3200 to 3500	(m) free groups of silanol (s) associated groups of silanol
$\equiv SiH$	2100 to 2300	(s)
$\equiv SiCH_3$	1250 to 1280 860 to 750	(s) (s)
$\equiv SiC_6H_5$	1120 1590	(s) (m)
$\equiv Si–O–Si \equiv$	1130 to 1000	(s) broad, possibly different maxima
$\equiv Si–CH=CH_2$	1400	(s)

* s = strong; m = medium

the hydrogen evolved upon base-catalysed hydrolysis or alcoholysis. Silanol can be determined by measuring the methane evolved with methyl Grignard reagent; water is corrected for by reaction with calcium hydride, which, unless specially prepared, does not react with silanol. Water and silanol can also be determined separately by IR techniques.

Other wet analytical methods for determining silicone content or functional groups have been largely replaced by the methods described previously.

7 Literature

[1] Holleman-Wiberg, Lehrbuch der Anorganischen Chemie, Verlag Walter de Gruyter, 102, 2007, pp. 992–993
[2] Walter Noll, Chemie und Technologie der Silicone, Verlag Chemie GmbH, Weinheim, 1968, pp. 24–37
[3] Silicones – Chemistry and Technology, Vulkan-Verlag Essen, 1991, pp. 7–19
[4] Walter Noll, Chemie und Technologie der Silicone, Verlag Chemie GmbH, Weinheim, 1968, pp. 38–43
[5] F. S. Kipping, J. T. Abraham, J. Chem. Soc., 1944, p. 81
[6] Silicones – Chemistry and Technology, Vulkan-Verlag Essen, 1991, pp. 26–28
[7] Encyclopedia of Polymer Science and Engineering, John Wiley & Sons, Vol. 15, 1989, pp. 265–269
[8] E. Schamberg/G. Koerner, Goldschmidt informiert, 04/84, No. 63, pp. 49–56
[9] Encyclopedia of Polymer Science and Engineering, John Wiley & Sons, 1989, Vol. 15, pp. 269–270
[10] DE 1618836
[11] EP 1142929
[12] EP 0157318
[13] EP 0535599
[14] Stoye/Freitag, Lackharze, Carl Hanser Verlag, München – Wien, 1996, pp. 343/344
[15] US 2676182
[16] Silicones – Chemistry and Technology, Vulkan-Verlag Essen, 1991, pp. 28–30
[17] M. Shinohara, Polym. Prep. Am. Chem. Soc., Div. Polym. Chem. 15, 1974, pp. 72–75
[18] DE 2828990
[19] Stoye/Freitag, Lackharze, Carl Hanser Verlag, München – Wien, 1996, p. 341
[20] Stoye/Freitag, Lackharze, Carl Hanser Verlag, München – Wien, 1996, pp. 342/343
[21] D. Hinzmann, T. Klotzbach, S. Herrwerth, Ein starkes Netzwerk gegen Rost, Farbe und Lack, 2012/9, pp. 22–26
[22] H. Kittel, Lehrbuch der Lacke und Beschichtungen, Band 2, S. Hirzel Verlag Stuttgart, 1998, pp. 173–174
[23] US 4898772
[24] H. Kittel, Lehrbuch der Lacke und Beschichtungen, Band 2, S. Hirzel Verlag Stuttgart, 1998, pp. 155–156
[25] Ulrich Poth, Polyester und Alkydharze, Vincentz Network Hannover, 2005, p. 208
[26] Stoye/Freitag, Lackharze, Carl Hanser Verlag, München - Wien, 1996, pp. 348
[27] US 6528607
[28] DE 10 2005 026 523

W. Heilen, S. Herrwerth: Silicone Resins and their Combinations

ISBN: 978-3-86630-828-2

[29] US 2009/0281207
[30] D. Hinzmann, T. Klotzbach, S. Herrwerth, Silicone-epoxy hybrid binders – a strong network against rust, Paint India, 2012, November, pp. 83–87
[31] WO 96/16109
[32] EP 1849831
[33] US 2002156187
[34] EP 1086974
[35] US 2014/0058012
[36] EP 0338550
[37] WO 2004/067576
[38] EP 1247823
[39] EP 0006517
[40] US 6281321
[41] EP 0941290
[42] Welt der Farben, 5/2012, pp. 10–12
[43] US 5162426
[44] EP 0549643
[45] WO 92/11327
[46] WO 2006/042585
[47] EP 2676982
[48] EP 2641925
[49] US 8569438
[50] EP 1273640
[51] WO 2008/074489
[52] S. Struck et al., Radcure Coatings and Inks, PRA 1996, Paper 8
[53] US 5863966
[54] US 6288129
[55] EP 0168713
[56] DE 3710238
[57] DE 3810140
[58] DE 3820294
[59] US 6548568
[60] US 6268404
[61] US 6211322
[62] EP 0617094
[63] US 6187834
[64] US 6902816
[65] Encyclopedia of Polymer Science and Engineering, John Wiley & Sons, Vol. 15, pp. 273–286
[66] Eugene G. Rochow, Silicium und Silicone, Springer-Verlag Berlin Heidelberg, 1991, pp. 123–134
[67] Ullmann's Encyclopedia of Industrial Chemistry, Wiley – VCH Verlag Weinheim, Vol. 32, 2012, pp. 694–695
[68] http://www.chemiedidaktik.uni-wuppertal.de, Prof. D. M. W. Tausch, Didaktische Silicondokumentation

[69] Ullmann's Encyclopedia of Industrial Chemistry, Wiley – VCH Verlag Weinheim, Vol. 32, 2012, pp. 644–645
[70] US 3445420
[71] US 3344111
[72] WO 03/099909
[73] US 5567764
[74] WO 03029375
[75] US 2891920
[76] US 3294725
[77] WO 2010062674
[78] US 5817714
[79] WO 2006081978
[80] Markus Antonietti, Katharina Landfester, Progress in Polymer Science 2002, Vol. 27, pp. 689–757: Polyreactions in mini-emulsions
[81] Cosima Stubenrauch (Ed.), Microemulsions: Background, new concepts, applications, perspectives, John Wiley & Sons, 2008
[82] W. Heilen, Lehrbuch der Lacke und Beschichtungen, Vol. 3, S. Hirzel-Verlag, Leipzig, 2001, pp. 256
[83] WO 2013000592
[84] G. Feldmann-Krane et al., Tego-Journal, 1. Edn. Feburary 1999, Tego Chemie Service GmbH, Essen
[85] M. Ferenz, Innovative Developments for UV Silicone Release Coatings, PSI Yearbook 2004, July 2004
[86] E. Schamberg, Goldschmidt informiert, 01/86, No. 64, pp. 26–37
[87] W. Heilen, M. Priesch, Dispersions-Silikatsysteme, TAE Esslingen, Expert Verlag Renningen, 1995, pp. 459–476
[88] W. Heilen, Siliconharz – Fassadenfarben, VILF Vortrag, Krefeld 1990
[89] O. Wagner, Wässrige Siliconharz-Beschichtungssysteme, TAE Esslingen, Expert Verlag Renningen, 1997, pp. 113–164
[90] www.photokatalyse.fraunhofer.de, Fraunhofer Allianz Photokatalyse
[91] www.newscientist.com, J. Hogan, 02/2004, p. 23
[92] Eugene G. Rochow, Silicium und Silicone, Springer Verlag, 1991, pp. 137–148
[93] M. Hesse et al, Spektroskopische Methoden in der organischen Chemie Thieme Verlag Stuttgart, 1995, p. 101
[94] H.O. Kalinowski et al, 13C-NMR-Spektroskopie, Thieme Verlag Stuttgart, 1984, pp. 14–43
[95] Ullmann's Encyclopedia of Industrial Chemistry, Fifth Edition, Wiley – VCH Verlag Weinheim, Vol. A24, pp. 86/87
[96] I. Jussofie, Analytisches Labor der Goldschmidt GmbH Essen
[97] Encyclopedia of Polymer Science and Engineering, John Wiley & Sons, Vol. 15, pp. 291–293

Authors

Wernfried Heilen studied chemistry at the University of Applied Chemistry in Krefeld, Germany. He started his career with Wülfing (PPG) in product development for the coatings industry and then moved to Byk Chemie where he was Product Manager responsible for various product groups. In 1983, he became Head of Technical Service in the Additives and later Silicone Resin areas for Goldschmidt. Since 2001, he has been Vice President (Technical Marketing) at Evonik Industries AG. He has many publications to his credit and is the holder of numerous patents.

Dr. Sascha Herrwerth was born 1972 in Hamburg, Germany. He studied chemistry at the Ruprecht-Karls-Universität Heidelberg and at the University of Bristol. He obtained his Ph.D. in Heidelberg in 2002 under the supervision of Professor Michael Grunze for work on "Oligoether terminated self-assembled monolayers on gold and silver". In 2003 he joined Evonik Industries AG where he worked in different functionalities. Since 2011 he is heading the Technical Service Department Speciality Resins.

Index

W. Heilen, S. Herrwerth: Silicone Resins and their Combinations

ISBN: 978-3-86630-828-2

E

F

G

H

I

L

M

N

O

P

Q

R

S

T

U

W

Y

www.ingramcontent.com/pod-product-compliance
Lightning Source LLC
LaVergne TN
LVHW021726021026
849161LV00004BA/734

* 9 7 8 3 8 6 6 3 0 8 2 8 2 *